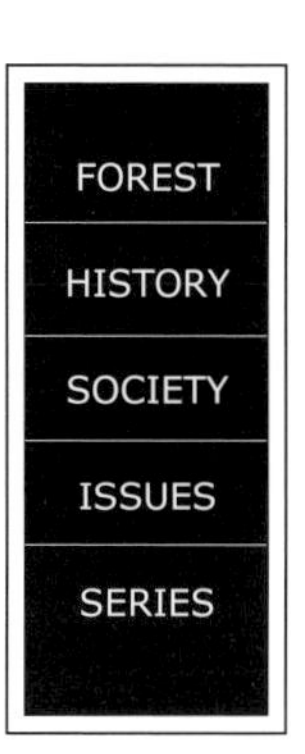

AMERICAN FORESTS

A HISTORY OF RESILIENCEY AND RECOVERY

DOUGLAS W. MacCLEERY

The Forest History Society is a nonprofit, educational institution dedicated to the advancement of historical understanding of human interaction with the forest environment. The Society was established in 1946. Interpretations and conclusions in FHS publications are those of the authors; the Society takes responsibility for the selection of topics, the competence of the authors, and their freedom of inquiry.

This edition of *American Forests* was sponsored by the Lynn W. Day Endowment for Publications in Forest History with support from the Reed-Henry Fund at the Seattle Foundation, MeadWestvaco Corporation, and Starker Forests Inc.

Forest History Society
701 Wm. Vickers Avenue
Durham, North Carolina 27701
919 682-9319
www.foresthistory.org

Cover photo by Bill Lea.
First published in 1992 by the USDA Forest Service, FS-540, 1992
Second printing, with revisions, 1993 by the Forest History Society
Third printing, 1994 by the Forest History Society
Fourth printing, with revisions, 1996 by the Forest History Society
Revised Edition, 2002 Forest History Society

Library of Congress Cataloging-in-Publication Data

MacCleery, Douglas W.
American forests: a history of resiliency and recovery / Douglas W. MacCleery.
p. cm. — (Forest History Society issues series)
Cover title: American forests.
ISBN 0-89030-048-8
1. Forests and forestry—United States—History. I. Title.
II. Title: American Forests. III. Series.
SD143.M32 1992
333.75'0973—dc20 92-29771
CIP

Forest History Society Issues Series

The Forest History Society was founded in 1946. Since that time the Society, through its research, reference, and publication programs, has advanced forest and conservation history scholarship. At the same time, it has translated that scholarship into formats useful for people with policy and management responsibilities. For more than five decades the Society has worked to demonstrate history's significant utility.

The Forest History Society Issues Series is one of the Society's most explicit contributions to history's utility. The Society selects issues of importance today that also have significant historical dimensions. Then we invite authors of demonstrated knowledge to examine an issue and synthesize its substantial literature, while keeping the general reader in mind.

The final and most important step is making these authoritative overviews available. Toward that end, each booklet is distributed to people with management, education, policy, or legislative responsibilities who will benefit from a deepened understanding of how a particular issue began and evolved.

The Issues Series—like its Forest History Society sponsor—is non-advocacy. The series aims to present a balanced rendition of often contentious issues. Although all views are aired, the focus is on consensus. The pages that follow document the resilience of American forests and establish a baseline for discussion. Many of today's debates hinge on "how much there is" and "how much there was." The Society believes that this work is of great value to those who make decisions and articulate policy.

The Society gratefully acknowledges financial support from the USDA Forest Service and the American Forest Foundation for this series title by Douglas MacCleery. Weyerhaeuser Company, Westvaco, and Champion International Corporation supported the revised fourth printing. The Lynn W. Day Endowment for Publications in Forest History sponsored the significantly revised and updated fifth printing, which was also supported by the Reed-Henry Fund at the Seattle Foundation, MeadWestvaco Corporation, and Starker Forests Inc.

Contents

Figures

Overview

Forests are resilient. It is a tribute to this inherent quality of American forests and to the success of the policies that were put in place in response to public concerns that forest conditions over much of the United States have improved dramatically since 1900. The following snapshot compares the forest situation as it was in 1900 with the way it is today:

- Following two centuries of decline, the area of forest land has stabilized (see figure 2). Today the United States has about the same forest area as in 1920.

- The area consumed by wildfire each year has fallen 90 percent; it was between 20 million and 50 million acres in the early 1900s and is between 2 million and 5 million acres today (see figure 5).

- Nationally, the average standing wood volume per acre in U.S. forests is about one-third greater today than in 1953; in the East, average volume per acre has almost doubled (see figure 14).

- Populations of white-tailed deer, wild turkey, elk, pronghorns, and many other wildlife species have increased dramatically (see figures 6, 7, 8, and 9). But some species, especially some having specialized habitat conditions, remain the cause for concern.

- Tree planting on all forest land rose dramatically after World War II, reaching record levels in the 1980s. Many private forest lands are now actively managed for tree growing and other values and uses (see figure 10).

- The tens of millions of acres of cutovers, or "stumplands," that existed in 1900 have long since been reforested. Many of these areas today are mature forests. Others have been harvested a second time, and the cycle of regeneration to young forests has started again.

- Eastern forests have staged a major comeback (see figure 12).

- Forest growth nationally has exceeded harvest since the 1940s. By 1997 forest growth exceeded harvest by 42 percent and the volume of forest growth was 380 percent greater than it had been in 1920 (see figure 15).

- Recreational use on national forests and other public and private forest lands has increased manyfold (see figure 16).

- The efficiency of wood utilization has improved substantially since 1900. Much less material is left in the woods, many sawmills produce more than double the usable lumber and other products per log input they did in 1900, engineering standards and designs have reduced the volume of wood used per square foot of building space, and preservative treatments have substantially extended the service life of wood. These efficiencies have reduced by millions of acres the area of annual harvest that otherwise might have occurred.

- American society in the 20th century changed from rural and agrarian to urban and industrialized. This has caused a shift in the mix of uses and values the public seeks from its forests (particularly its public forests). Increased demands for recreation and protection of biodiversity are driving forest management in some areas. In spite of this shift, today's urbanized nation is also placing record demands on its forests for timber production. We are no less dependent on the products of forests and fields today than were the subsistence farmers of America's past (see figure 18).

Introduction

Forests are a key element in the broad sweep of United States history. The forest landscape has changed greatly over time, as has public concern for trees, water, and wildlife. The conservation movement of the early 20th century and the policy changes that resulted from that movement have been leading factors affecting the forests of today.

The single most important event in the evolution of the modern American landscape was the clearing of forests for agriculture, fuelwood, and building material.

People depended heavily on the products of the forest both in their personal lives and in the general economy. Wood was virtually the only fuel used in this country until the last half of the 19th century. Wood warmed people, cooked their food, produced iron, and drove locomotives, steamboats, and stationary engines. People used lumber, timbers, and other structural products as the primary material for building houses, barns, fences, bridges, and even dams and locks. These wood products were essential to rural economies across the nation, as well as to industry, transportation, and the development of towns and cities.

Forests were also habitat for the wildlife that supplemented the diet of millions of Americans for centuries. However, even more important to the American diet was food produced on land cleared of its forests and employed for agricultural use. This was by far the primary cause of forest loss.

In the spiritual dimension, the forest, and the wildness it represented, also played an important role in the identity of the nation. This was expressed in the writings of Henry David Thoreau, Ralph Waldo Emerson, George Perkins Marsh, and others, and was first evidenced politically during the late 1800s by efforts to address concerns over the decline of wildlife populations and the loss of forests. There is no question that without its forests, the United States of America would have had a decidedly different history, and would be a decidedly different place than it is today.

The Forest Prior to European Settlement

The original forest covered about 1 billion acres, or about half of the U.S. land area (including Alaska). About three-quarters of that forest covered the eastern third of the country. Today we have 747 million acres of forest, about 70 percent of the original forest. About 350 million acres has been converted to other uses since 1600, primarily to agricultural lands (see figure 1).

Forests remained the dominant feature of the landscape in eastern North America for centuries after initial settlement. In 1796, almost two centuries after the first European settlements, a French naturalist visiting the new American nation wrote, "The most striking feature [of the country] is an almost universal forest, starting at the Atlantic and thickening and enlarging to the heart of the country." He said that in his travels to America's interior he "scarcely passed, for three miles together through a tract of unwooded or cleared land."

This country's forest was and is magnificent and diverse. East of the Mississippi River, deciduous and coniferous forests blanket New England; open and sunlit pineries cover the southern coastal plain and Piedmont; remarkably varied and productive central hardwood forests extend from the central and southern Appalachians through the Ohio Valley and central Midwest; extensive pine and oak woodlands of the prairie fringe grow in Texas, Missouri, Indiana, Illinois, and Ohio; and the cool hardwood and coniferous boreal forests shade the northern Lake States.

West of the Mississippi River, rainfall diminishes, and forests and woodlands give way to treeless prairies and deserts. But in mountainous areas of the West where rainfall is sufficient, and along the Pacific Coast, extensive forests flourish. Fire-maintained lodgepole pine, ponderosa pine, and mixed-conifer forests cover the slopes of the Rocky Mountains and areas east of the Cascade and Sierra ranges in Washington, Oregon, and California. The most magnificent western forests grow along the rain-drenched and fog-shrouded coasts of the Far West, where coast redwood and Douglas-fir, Sitka spruce, and hemlock form vast, cathedral-like stands.

Besides being impressed by North America's seemingly boundless forests, early explorers were astounded by the abundance and variety of its wildlife. They reported prolific numbers of large mammals in the eastern forests, such as white-tailed deer, elk, moose, and bison. They also spoke of the incalculable numbers and remarkable variety of bird life: game birds, such as ruffed grouse, wild turkey, and heath hens, and waterfowl, including ducks, geese, herons, egrets, and ibises. The most abundant bird on the North American continent was the passenger pigeon, which darkened the sky in numbers that seem incredible today.

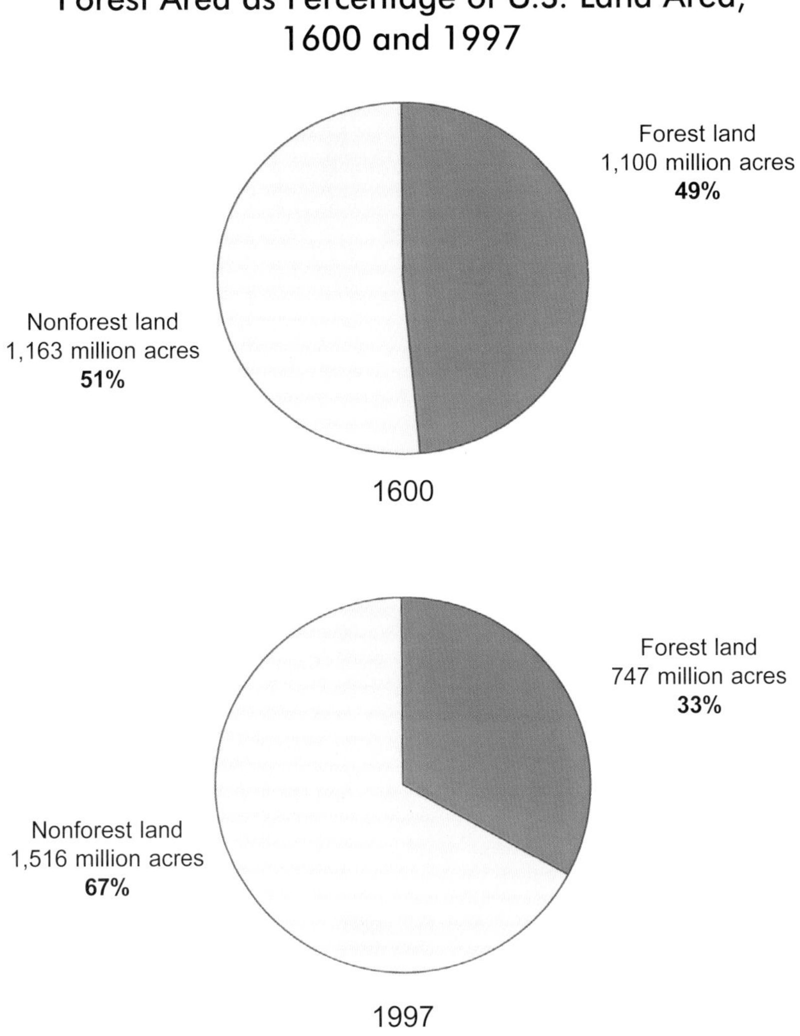

Figure 1. Forests covered slightly more than one billion acres, or somewhat less than half of the nation's land area, in 1600. Some of this land gave way to agriculture, reducing the acreage of the U.S. forest land base. About 350 million acres of forest has been converted to other uses since 1600—primarily to agriculture. Today about a third of the nation is forested, approximately two-thirds of the area that was forested in 1600.

Source: Forest Resources of the U.S., 1997. GTR-NC-219, USDA Forest Service, 2001.

Native Peoples' Effect on American Forests

One popular myth is that prior to European contact, America was dominated by impenetrable, relatively uniform ancient forests that cloaked the landscape in a long-term, static balance with the environment. The reality was far different. Pre-European settlement forests were exceedingly dynamic, shaped by myriad natural and human influences, disturbances, and catastrophic events that had a profound effect on the age and species mix of both plants and animals. The diversity of forest conditions that resulted from these influences was a major factor in creating the wildlife variety and abundance that so impressed early European settlers.

Forests in the country's East and West were not pristine. They were often strongly influenced by native peoples. In the eastern forests, humans lived in fixed villages and practiced a maize-based agriculture. Domesticated crops commonly accounted for half or more of their diet, with the remainder provided by wild berries, nuts, fruits, and game gathered from the adjacent forest.

Although pre-European settlement population figures are constantly debated and revised, what is truly significant is the impact of these peoples on the land. In addition to areas largely cleared of trees for crops, thousands of additional acres around each village were burned periodically to improve game habitat, facilitate travel, reduce insect pests, remove cover for potential enemies, enhance conditions for berries, and drive game. For example, in New England it was reported that the native peoples underburned the woods twice a year, in the spring and in the fall. Roger Williams (1604–1684), founder of the New England colony of Rhode Island in the 1640s, wrote that "this burning of the Wood to them they count a Benefit, both for destroying of vermin, and keeping downe the Weeds and thickets."

Early observers reported prolific numbers of animals along forest edges and openings, indicating a forest in which natural or human-induced disturbance was common. Even elk and bison, normally associated with the western prairies, were present in the eastern forests. In the early 1600s, bison were found grazing along the Potomac River in what is now Virginia and Maryland. Bison were reported in Massachusetts. The presence of these grazing animals indicates abundant grass and forbs that could only have been created by fire.

The South was dominated by fire-created forests, such as longleaf pine savannas on the Coastal Plain and Piedmont. The hardwood forests of the Appalachian Mountains were also burned frequently by native peoples. Virginia's Shenandoah Valley—the area between the Blue Ridge Mountains and the Alleghenies—was one vast grass prairie. Native peoples burned the area annually.

During the late 1580s the Indians of the village of Secoton, near Sir Walter Raleigh's colony of Roanoke in present-day North Carolina, raised abundant corn crops as well as some sunflowers and squash. This engraving by Theodore DeBry is after a watercolor by John White, the original leader of the colony. *Source:* Library of Congress.

On the western fringe of the eastern forest, fire-dominated forests, such as oak and pine savannas, covered tens of millions of acres. These forests were heavily influenced by fires sweeping off the prairies. Fire-created prairies extended into Ohio, Pennsylvania, and even western New York. Evidence of the dominant role fire played in these forests is demonstrated by the fact that when farmers finally began to move out onto the prairies and reduced wildfires, millions of acres of open oak savannas and even treeless areas to the east of these farms became dense forests and woodlands within two decades.

Today, with rising interest in protecting more forests in their "natural" condition, the complex pre-European settlement history raises technical and policy questions over whether and how to allow wildfire to assume its natural role in these areas, and whether to seek to replicate pre-European settlement human influences. It is difficult and sometimes impossible to distinguish natural from human-caused influences in pre-European settlement forests: North American forests have been both occupied and influenced by humans from the time these forests advanced north behind the retreating continental glaciers more than 8,000 years ago.

Changes Brought to the New World, 1500–1785

European settlement ushered in a vast increase in the impact of humans on the forest. The abundance of land and resources and the scarcity of labor was a defining difference between America and Europe, where the situation was reversed. This difference was profound, affecting everything from the way resources were utilized to the type of stewardship applied to the land, as well as the adoption of slavery.

Both fishing and fur trading thrived before permanent settlements were established in what is now the United States. Fur trading based on beaver, otter, lynx, and many other forest-dwelling animals was one of North America's first industries, and its success depended on the active involvement of native peoples as hunters and trappers. The astoundingly productive Atlantic fishery formed the foundation of a lucrative industry that began along the Atlantic coast in the 1500s.

Lumber was also one of the first exports from the New World. In 1621, only a year after the *Mayflower* arrived, the Pilgrims sent the ship *Fortune* back to England "laden with good clapboard as full as she could stow." Soon the colonies became the source of white pine ship masts, oak planking, and cedar timbers, upon which the English navy depended. The forests of England had long since been depleted of ship-grade material; supplies from

Castle Creek in the Black Hills during the 1870s (top) and following a century of fire exclusion (bottom) that allowed a forest to appear. *Source:* South Dakota Agricultural Experiment Station photo.

the Baltic States, where England was then obtaining its masts and timbers, were of lower quality, expensive, and subject to political disruption. By the middle of the 1600s, the colonists had established a booming business in ship masts, naval stores (such as pitch and turpentine), timbers, and other forest products.

Early European colonists viewed the seemingly endless forest as a mixed blessing. On one hand it provided an abundant and available source of fuel and building materials. It yielded game that for decades after settlement remained an important food source. But the forest was also habitat for wolves, eastern panthers, and other predators that found colonial livestock easy prey and against which the colonists waged unrelenting war. It provided cover for sometimes hostile Indians. But most importantly, it occupied potential cropland that could be liberated only after intensive and back-breaking labor using hand tools.

For the first three centuries of U.S. history, most Americans were farmers. Ninety-five percent of the people lived on the land in 1800. Except for a relatively few people engaged in plantation agriculture in the South, most were subsistence farmers. From this perspective, the predominant view that emerged in the early 1600s—and continued for almost 300 years—was that the forest was both inexhaustible and an obstacle to the preferred agricultural use of the land.

The colonists cleared the forests using techniques learned from the native inhabitants, but with the substantial advantage of iron tools and draft animals. Initially, white settlers sought abandoned Indian fields, which required less labor to clear than did a mature forest. Clearing forests was extremely laborious and time consuming. About one man-month of effort was required for each acre of mature forest cleared (assuming the axman was strong and healthy). Trees were either felled with an ax and removed before planting, or they were killed in place by girdling (removing the bark in a band around the tree) and left standing. In both cases, fire helped clear the undergrowth.

The settlers planted crops borrowed from the Indians—corn, squash, tobacco, beans, and pumpkins. Other crops first domesticated by native peoples include both white and sweet potatoes, tomatoes, blue grapes, peanuts, sunflowers, both sweet and chili peppers, strawberries, cocoa beans (chocolate), vanilla, avocados, pineapple, cassava, cotton, and gourds. American agriculture still relies heavily on native crops. Today almost 60 percent of the value of U.S. crop production consists of plants first domesticated by native peoples.

The most significant difference between European and native agriculture was that the Europeans possessed livestock and draft animals. Within a few years after settlements were established in an area, the numbers of livestock increased dramatically. In 1634 the Massachusetts Bay Colony had a population of 4,000 people, 1,500 cattle, 4,000 goats, and "swine innumerable."

Because labor was scarce, the common European practice of herding livestock was generally not practiced. Instead, hogs, cattle, and other livestock were turned untended into the woods, which meant that fences were needed to keep them out of crops and gardens.

Next to clearing the forest and constructing farm buildings, the most labor-intensive activity in creating a farm was building fences. One observer wrote that "it is inconceivable the cost and care which a single large farm requires in that single item." A square 40-acre field enclosed by a wooden

Fencing. Hand-split rails are stacked in zigzag fashion. Although these rail fences have a firm place in American pioneer folklore because of Abraham Lincoln's well-known youthful chores, they consumed large quantities of wood and were impractical when the westward-moving frontier reached the prairie region. *Source:* Forest History Society Lantern Slide Collection.

zigzag fence required about 8,000 fence rails. An average farmer could split 50 to 100 rails in a day.

Until woven wire and barbed wire were introduced in the latter half of the 19th century, farm fences were made of wood or stone. The volume of wood used in farm fencing substantially exceeded that of lumber until the 1840s. By 1850 there were about 3.2 million miles of wooden fence in the United States, enough to encircle the earth 120 times.

The abundant forests also provided European settlers a level of physical comfort in winter unknown in the forest-depleted Old World. In 1650 an English visitor, Francis Higgins, wrote that a "poor servant here...may afford to give more wood for Timber and Fire...then many Noble men in England can afford to do."

Such comfort came at a price. In the late 1700s, about two-thirds of the volume of wood removed from the forest was used for energy. Wood provided virtually all of the energy consumed in the United States. Heating and cooking were done in inefficient fireplaces. It was not uncommon for a single household to consume 15 to 40 cords of wood annually. Thus in a single year more wood went up the chimney in smoke than had been used to build the house that was being heated. The average per capita consumption of fuelwood was about 4.5 cords per year throughout the colonial period.

Wood for fuel went far beyond meeting domestic needs for heating and cooking. It was also used to produce iron and other metals critical to the country's economy. Virtually all iron produced in America throughout the 18th century was smelted using wood charcoal. The reason was clear: wood was abundant, the technology was simple, and it could be done in a small operation. Blacksmiths found charcoal iron malleable and easy to shape into a variety of tools and other iron products.

Nearly every American colony had a number of iron-making furnaces. By the late 1700s many individual ironworks were producing 1,000 tons or more of iron per year. Thus the impact on the local forest was significant. A 1,000-ton ironworks required between 20,000 and 30,000 acres of forest to sustain itself over time.

As settlers continued to clear forests for farms, firewood, and energy production, wildlife populations dropped dramatically. Even before the middle of the 1700s, many game animals and furbearers, such as deer, eastern elk, wild turkey, and beaver, were becoming scarce in many areas. Trappers practically eliminated beaver east of the Appalachians by 1700. These areas would not see the beaver's return for almost two and a half

Charcoal making in 1900 used methods that had gone unchanged for centuries. The wood is stacked in the shape of a cone, covered with earth, and then ignited and carefully tended. *Source:* Forest History Society photo.

centuries. Wild turkeys were considered rare in many locations by 1670, and the bison was gone from the East before the Revolution of 1776.

This decline in game species was not primarily the result of habitat loss. On the contrary, habitat conditions in many parts of the colonies would have been ideal for deer, wild turkey, and beaver. The problem was in the social and property arrangements designed for the taking of desired species. Because wildlife crosses property lines at will, and ownership to it does not pass until it is killed, normal property arrangements do not work. Individuals therefore have little incentive to conserve game if their neighbors do not, because in economic terms they suffer a known loss with little perceived benefit. Today this difficulty in conserving common property assets is called the "tragedy of the commons."

In Europe, the "commons" problem was effectively, if undemocratically, dealt with by the nobility, who decreed that all wild game was the property of the crown and any commoner caught taking it would be severely punished. But this institutional arrangement did not cross the Atlantic. Initially it was not perceived as a problem because of the small human population and abundant wildlife. It soon became apparent that some form of regulation was needed, but it would be well into the 20th century before the country could muster the social will to institute and enforce effective game regulations.

Westward Expansion and Eastern Industrial Growth

The massive Louisiana Purchase of 1803 doubled the nation's land area, and by 1850 the land base for the forty-eight contiguous states was in place. Acquired land was added to the public domain. Throughout most of the 19th century, the government viewed it as in the national interest to rapidly transfer most public domain lands to private ownership. The work became the largest and longest-lasting privatization effort in the history of the world. The increase in land transfers reflected a parallel increase in population.

It had taken the colonies a century and a half to reach a population of 3 million people. However, in the sixty-five-year period between 1785 and 1850, the U.S. population increased more than seven times, to 23.3 million people.

Since it required an average of about 3 acres of cropland to support each person, the area of cropland grew at about the same rate as the population. By 1850 the total cropland area, which had been about 20 million acres in 1800, had grown to 76 million acres. Clearing for pasture and hay land substantially added to that figure. Farmers and settlers carved much of this agricultural land out of the forest (see figure 2).

Expansion of population and industry put increasing pressure on U.S. forests, both east and west of the Appalachians. Water-powered mills operated next to New England rivers and streams. Farms in New England, which had previously functioned at subsistence levels, prospered as they provided for the communities growing up next to these mills. Farmers cleared large areas to pasture sheep that would provide wool for the textile industry; beef cattle provided meat and hides to growing areas in the East, both for domestic use and for export. In the South, forests were cleared for cotton, tobacco, and other crops.

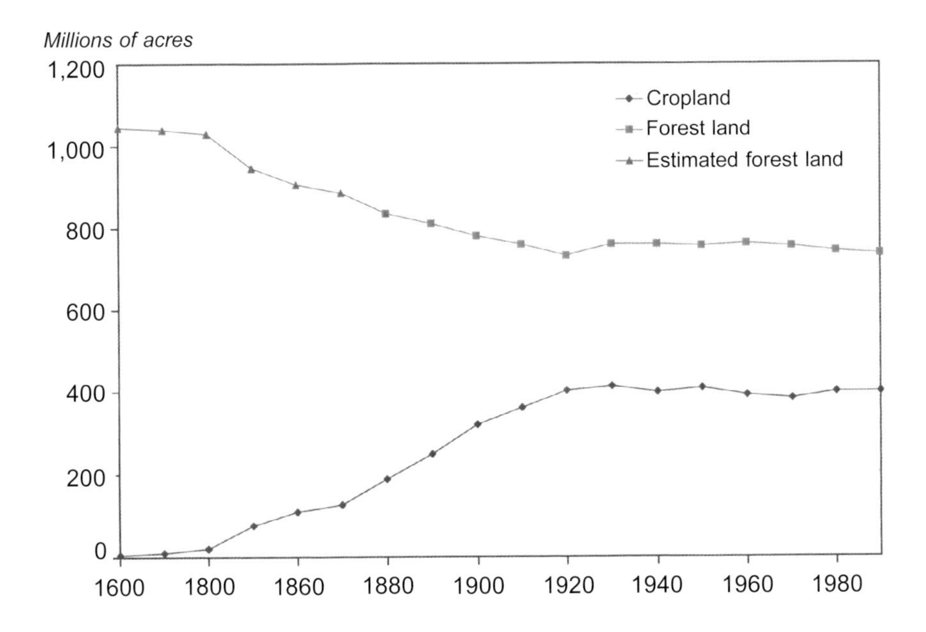

Figure 2. The nation's forest land area is about the same size today as it was in 1920, when the acreage devoted to cropland stabilized. Two factors contributed to this stabilization. First, as horses, mules, and other draft animals were replaced by farm tractors and motor vehicles, cropland formerly used to feed draft animals was freed for use in human food production. Second, after 1930 agricultural productivity began to improve because of genetically improved crops, irrigation, and increasing use of fertilizers. Today U.S. farmers produce crop yields that are five times greater per acre than those produced in 1920. Note: Scale from 1600 to 1860 is compressed on the graph.

Sources: J. Fedkiw, GTR RM-175, USDA-Forest Service, 9/89; Forest Resources of the United States, 1997; GTR-NC-219, USDA-Forest Service, 2001.

Use of the Forest for Fuel

The volume of wood used in 1850 was almost six times the volume of fifty years earlier. By mid-century, wood still supplied more than 90 percent of the nation's heat energy needs; domestic heating and cooking accounted for the largest use of wood fuel (see figure 3).

The increasing scarcity and expense of fuelwood spurred innovations in the form of cast-iron wood stoves, which were four to six times more efficient in the use of wood than fireplaces. In the fifty-five years between 1790 and 1845, the U.S. Patent Office issued more patents for stoves (more than 800) than for any other object. But in spite of their obvious advantages, adoption of wood stoves was gradual, occurring first in towns, where wood was expensive as well as difficult to store because of its bulk. Fireplaces continued to predominate for cooking and heating in rural areas well into the mid-1800s.

While increased use of wood stoves began to reduce the per capita consumption of fuelwood for domestic purposes, increases in industrial uses of fuelwood (including ironmaking and fuel for growing numbers of

Southern Appalachian farmers eked a living from small clearings in the forest. USDA Forest Service photo.

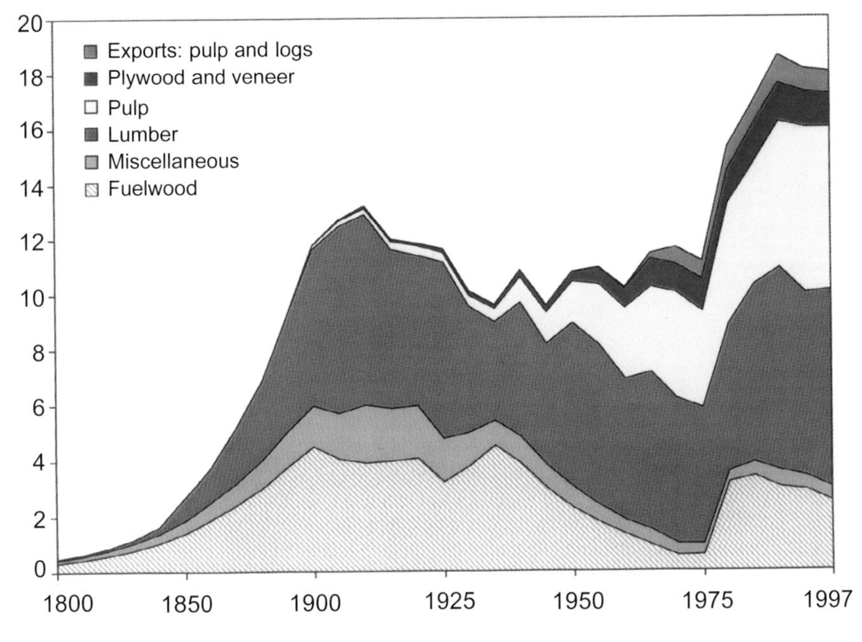

Figure 3. During the first half of the 19th century, domestic output of forest products rose at the rate of population growth. Heating and cooking was the largest use of wood during this period, averaging from one-half to two-thirds of total wood use. In 1850 wood provided more than 90 percent of the nation's energy. After 1900, fossil fuels largely replaced wood fuels, and wood substitutes, such as steel and concrete, replaced wood in some structural applications. In addition, there were significant gains in efficiency in the utilization of wood in logging operations, at the mill, and in end product uses. The rising real price of wood encouraged such changes. The price of timber, adjusted for inflation, had risen steadily since 1800, increasing about five times during the century. The real prices of most materials that competed with wood were steady or declining during this period. Note: Scale from 1800 to 1900 is compressed on the graph.

Sources: Sedjo, Roger A. 1991. "Forest Resources: Resilient and Serviceable" in *America's Renewable Resources: Historic Trends and Current Challenges*, D. K. Frederick and Roger A. Sedjo, eds., Washington, D.C.: Resources for the Future; and "U.S. Timber Production, Trade, Consumption & Price Statistics: 1965–97," FPL-GTR-116, USDA Forest Service.

steamboats and railroad locomotives) offset these gains. Consequently, the per capita consumption of fuelwood for all purposes remained at more than four cords per year until the late 1800s. Because the population expanded more than fourteen times between 1800 and 1900, and per capita consumption of fuelwood remained constant, there was increasing pressure on many forest areas. This led to forest depletion in some areas and local shortages. A traveler reported that on the 240-mile journey between New York and Boston in the early 1800s he passed through less than 20 miles of woodland, scattered in four or five dozen separate parcels.

Fuelwood remained the primary product of the forest until the 1880s, when the volume of lumber finally exceeded it. Although the volume of wood used for energy continued to increase until 1900, it supplied a progressively lower proportion of U.S. energy needs. As the country began to turn to coal, and later to oil for its energy needs, wood dropped from supplying more than 90 percent of the nation's energy in 1850 to 75 percent in 1870 to about 10 percent in 1920 (most of which was consumed by farm families). Yet even the move to coal increased the demand for wood in the form of millions of mine props to support deep mining operations in the mountains. Today wood energy supplies about 3 percent of U.S. energy needs, two-thirds of which is produced in industrial processes.

Ironmaking

Production of charcoal iron continued to increase after 1800. In 1810 England had not one charcoal iron furnace; all were coal- or coke-fired. In the United States at that time, there were no coke-fired furnaces.

In the 1850s the tonnage of coke iron produced finally exceeded that of charcoal iron. Even so, charcoal iron production continued to rise until 1900. Charcoal iron continued to be used after 1900 for specialty products. Because of its special properties, some early car makers specified it for engine blocks. The last charcoal-fired iron furnace finally shut down in 1945.

Transportation

By the early 1800s, the United States was one of the largest nations in the world. The transportation system more than anything else tied the disparate and often quarreling states together. America's forests figured heavily in building this system.

The nation's first highways were its rivers, where wooden keelboats and, after 1830, steamboats transported goods. Steamboats were made of wood and, until the Civil War, used wood for fuel. In 1840 almost 900,000 cords of wood were sold for steamboat fuel, representing one-fifth of all fuelwood sold.

Following steamboats came railroads. After 1850 railroads began expanding rapidly, linking growing cities and providing access to market for agricultural and forest products. Although called the "iron road," railroads used far more wood than iron. Except for the engine and rails, railroads were made of wood: cars were wood, ties were wood, the fuel was wood, the bridges and trestles were wood, and station houses, fences, and telegraph poles were wood.

The number of miles of U.S. railroads increased from less than 10,000 miles to more than 350,000 miles between 1850 and 1910. By the late 1800s, railroads accounted for 20 to 25 percent of the country's total consumption of timber.

By far the most significant railroad use of wood was for crossties. Each mile of track required more than 2,500 ties. Crossties were not treated with preservatives until after 1900, so because of their rapid deterioration in contact with the ground, they had to be replaced every five to seven years. Given the miles of track in 1910, that would be equivalent to replacing the ties on some 50,000 miles of track annually. Just replacing railroad ties on a sustained basis required between 15 million and 20 million acres of forest land in 1900.

Population and Agricultural Growth

The five decades from 1850 to 1900 witnessed an unprecedented demand for and impact on the nation's natural resources. Its forests, croplands, grasslands, and wildlife populations and habitats felt increasing pressure. Rising population and increasing urbanization drove this demand.

Between 1850 and 1900, the population tripled, from 23 million to 76 million. Even the bloody decade of the Civil War showed a 27 percent increase. Immigration added to this population growth, amounting to 32 percent of the nation's growth during the last half of the 19th century.

Increased industrialization might logically reduce a nation's demands on its forests. Coal replaced wood fuels, and objects formerly made of wood, such as buildings, fences, bridges, nails, and machinery, were increasingly

Top: Railroad bridges and trestles were constructed of pilings and large timbers. *Source:* Forest History Society photo. Above: Railroads consumed vast quantities of wood, as these stacked crossties show. *Source:* B.C. Forest Service photo.

made of brick, iron, steel, and other materials. In fact, industrialization in Europe was partly a response to diminishing wood supplies. In America, however, other factors were the driving force, including improving transportation systems and adoption of European industrial technologies.

In the second half of the 19th century, extensive land was cleared for farming. During this period, while the U.S. population tripled, the total area of cropland increased by more than four times, from 76 million to 319 million acres.

For every person added to the U.S. population during the 19th century, farmers put another 3 to 4 acres under the plow. Except for the decade of the Civil War, the increase in the area cleared for cropland paralleled the increase in U.S. population (see figure 4). Between 1850 and 1910, farmers cleared about 190 million acres of forest for crops and pasture, an amount greater than the total over the previous 250 years of settlement. In fact, during the sixty years between 1850 and 1910, the nation's farmers cleared at an average rate of 13.5 square miles of forest per day.

All sections of the country contributed to forest clearing for agriculture between 1850 and 1910, with about 44 million acres (23 percent) occurring on the Pacific Coast and in the Southwest; and 146 million acres (76 percent) in the East and South. Ohio was typical of farm clearing in the Midwest. In 1800 about 96 percent of the state was covered with hardwood forests, with the remainder in grass prairies probably maintained by Indian-set fires. Fifty years later, forest still covered about 60 percent of the state, but by 1900 forests covered only 25 percent of the state. In the productive farm country of the western half of the state, forest cover in many areas was reduced to 4 percent of the land.

It was well into the 20th century before gains in per-acre agricultural productivity were made. Such gains were essential to reduce the rate of land clearing to feed a growing population, and were a prerequisite to reducing pressures on the nation's remaining forests and wildlife habitats.

Expansion of Lumber Production

Throughout the first half of the 1800s, most sawmills were small-scale, two- to five-person operations. Census figures for 1840 indicate that the number of sawmills in the United States was 31,649, or an average of twenty-five mills for every county. The number was much higher in eastern counties than for areas newly settled, with some counties along the Atlantic seaboard having more than one hundred mills, and pioneer counties in the Midwest and South having fewer than ten mills.

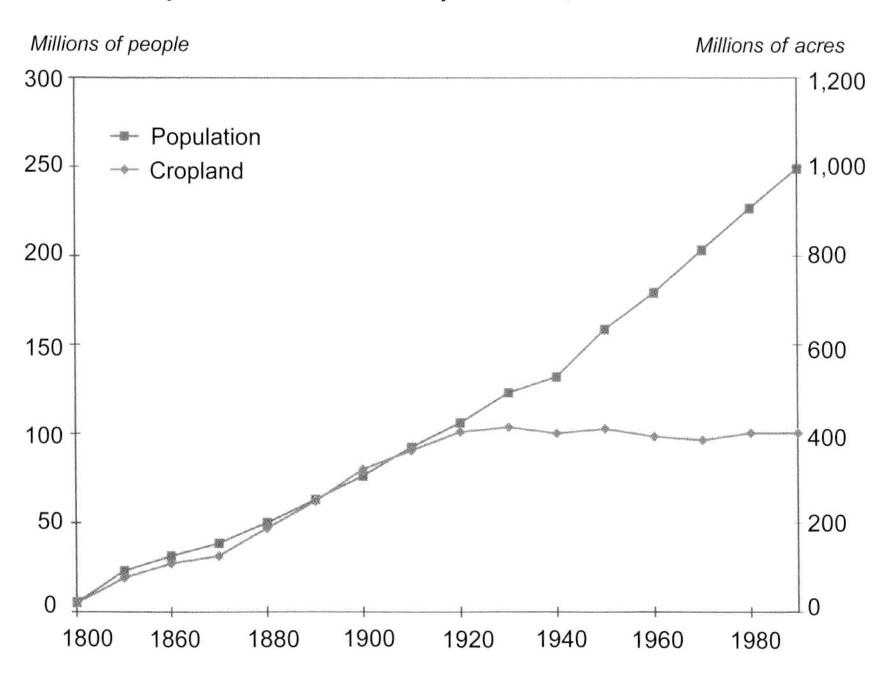

Figure 4. The U.S. population rose more than 14 times during the 19th century. Since farm productivity was not increasing on a per-acre basis, land clearing continued at about the rate of population growth.

Between 1850 and 1900, the U.S. population increased more than threefold, from 23 million to 76 million people, while the area of cropland increased fourfold, from 76 million to 319 million acres. For every person added to the U.S. population during the 19th century, farmers were putting another 3 to 4 acres of cropland under the plow. The area of pasture and hayland increased even more than that of cropland.

In the 1920s, the three-century-long conversion of U.S. forests to farmland largely halted. Today, the nation has about the same area of cropland as in 1920, even though the U.S. population has increased more than 2.6 times, from 106 million to 286 million, and U.S. farmers also feed the equivalent of more than 100 million people in other lands.

The vast improvement in agricultural productivity, which made possible the stabilization of cropland area, is a truly remarkable accomplishment that has benefited American forests.

Note: Scale from 1800 to 1860 is compressed on the graph.

Sources: J. Fedkiw, GTR RM-175, USDA-Forest Service, 9/89; and U.S. Bureau of Census Figures.

Until 1850 small country sawmills handled most of the nation's wood needs—either as a result of farm clearing or from farm woodlots. Often it was the farmers themselves who cut the timber and cordwood and operated the sawmills.

Although farmers cleared at record levels after 1850, the process generated too little wood to meet rapidly increasing demand. The location of the clearing was also a problem; rural communities could meet their wood needs with local production, but the large quantities of lumber and other wood products that cities demanded required new arrangements for manufacturing and transporting forest products. Also away from the city and away from the forests, prairie farmers west of the Mississippi began to demand large quantities of wood for houses, barns, fences, outbuildings, and fuel. As the physical distance between consumers and forests grew, logging and sawmilling increasingly became large-scale, industrial operations.

Lumber production increased dramatically, rising more than eight times between 1850 and 1910, from 5.4 billion board feet to 44.5 billion board feet annually, a rate more than double the rate of population growth.

Farmers and loggers burned limbs, tops, and other logging debris, believing that the logged areas could be converted to cropland or improved pasture. These uncontrolled slash fires burned nearly continuously and under some weather conditions resulted in massive wildfires that destroyed property and lives.

The South escaped much of the destructive postlogging fires that occurred in the North, perhaps because many of the native southern pine stands were of a more open type that had been maintained by frequent natural or Indian-set fires. Southern farmers continued the native practice of burning the woods, which reduced undergrowth and fuel buildup necessary for large wildfires.

Wildlife

The buffalo was one of the most dramatic examples of a large number of wildlife populations that by the last half of the 19th century had been severely diminished. By 1890 people had eliminated the white-tailed deer from much of its range east of the Mississippi, including all the New England states west of northern Maine, as well as Maryland, New Jersey, Ohio, Pennsylvania, and the Lake States except the extreme northern portions of Michigan, Minnesota, and Wisconsin.

No longer was wildlife abundant. Populations had declined because of unrestricted market hunting of all kinds of wildlife for food, furs, and feathers as well as habitat modification caused by farm clearing, logging, and extensive wildfires. Even many songbirds—such as robins and meadowlarks—were heavily hunted for food.

Emergence of a Call for Action

Before the end of the 19th century a growing number of people became concerned about what was happening to the nation's woodlands and wildlife. The combination of logging, massive wildfires, farm clearing, and wildlife depletion began to call into question the notion of the forest's inexhaustibility. Fears about future timber supplies combined with implications for increased flooding and watershed damage, declining wildlife populations, harm to the beauty of the American landscape, and even concerns about how forest clearing was affecting the climate itself. George Perkins Marsh raised concerns about the adverse effects farm clearing had on watersheds and other environmental values. His 1864 book, *Man and Nature*, became a catalyst for public concern. As early as 1865, Frederic Starr predicted an impending "national famine of wood"—a concern that would be raised frequently in the next few decades. Use of the term "famine" was apt, for wood in its various forms was among the most widespread and essential materials for both domestic use and industry.

The rapidity of change led to public concern as people in some areas watched the landscape, in forty or fifty years, lose 80 percent of its forests. At first the conservation movement was not organized. Groups with common interests moved more or less independently, seeking to achieve similar results through their efforts. Some of these groups began to set aside land in protective areas: Yosemite in California (1864); Yellowstone in Wyoming (1872); and the Adirondack Preserve in New York (1885). In 1891 Congress authorized the president to designate forest reserves from public domain lands but made no provision for their management. The forest reserves, unlike the park preserves, were generally not tied to the preservation of a nationally significant unique area. These reserve designations, which had grown to 44 million acres by 1897, generated considerable opposition because no one knew how such setasides would address society's need for water, forage, wood products, and other resources.

Scientists—including foresters like Bernhard E. Fernow and geologists like John Wesley Powell and Arnold Hague—prepared reports, wrote articles, and testified to Congress about the need to protect forested watersheds, water for irrigation, trees, forage, and wildlife. Citizen groups—the American Forestry Association and the Boone and Crockett Club—also advanced the cause. The result of these efforts was that Congress gave the forest reserves a management mandate (through the 1897 Organic Act) to "preserve and protect the forests," to "secure favorable conditions of water flows," and to "furnish a continuous supply of timber for the use and necessities of the citizens of the United States."

Congressional and presidential actions to reserve national forests from public domain lands signaled a shift away from three centuries of national policy that was designed to transfer public land to private ownership. By 1900 more than a billion acres of public domain lands, more than half the land area of the contiguous forty-eight states, had been transferred to private ownership.

The turn of the century also signaled a general change in how people viewed natural resources. It was becoming clear that the myth of forest and wildlife inexhaustibility was untenable and that the existing rates of forest and wildlife consumption were not sustainable. Although new approaches were called for, it was not clear what shape these approaches would take.

The idea of "conservation as wise use" emerged and received widespread public support under the dynamic advocacy of Gifford Pinchot and his friend and mentor, President Theodore Roosevelt. Pinchot and other conservation leaders were influenced strongly by Progressive Era thinking, which put great faith in science and the rational approach. Their view supported faith in efficiency but strong distrust of the "special interests" in politics. From this Progressive Era came the idea of conservation as the "wise use" of natural resources. Under this view, current use of resources should protect the basic productivity of the land and its ability to serve future generations.

Condition of Forests and Wildlife in 1900

The following snapshot of the condition of the nation's forests and wildlife in 1900 helps frame the natural resource situation that faced early conservation leaders:

- Wildfire commonly consumed 20 million to 50 million acres annually (an area the size of Virginia, West Virginia, Maryland, and Delaware combined).
- About 80 million acres of "cutovers" continued to be either idle or lacking desirable trees.
- The volume of timber cut nationally greatly exceeded that of forest growth.

Only stumps remained on cutover and abandoned forestland in northern Michigan at the beginning of the 20th century. *Source:* Forest History Society photo.

- There was no provision for reforestation. Aside from a few experimental programs, long-term forest management was not practiced.
- Wood was still relatively cheap; because of this, large quantities were left after logging, sawmills were inefficient, use of wood in buildings was based on custom rather than sound engineering, and huge volumes of wood simply rotted.
- Massive clearing of forest land for agriculture continued: in the last fifty years of the 19th century, forest cover in many areas east of the Mississippi had fallen from 70 percent to 20 percent or less. In the last decade of the century, America's farmers cleared forests at the average rate of 13.5 square miles per day. Much of this land included steep slopes that were highly erodible.
- Formerly abundant wildlife species were severely depleted or nearing extinction. Among them were white-tailed deer, wild turkey, pronghorn, moose, black bear, bighorn sheep, and bison. Furbearers,

especially beaver, had been eliminated from significant portions of their ranges. Waterfowl were severely affected, including wood ducks, Canada geese, and plumed wading birds (such as herons, egrets, ibises). The passenger pigeon, once the most abundant bird on the North American continent, was nearly extinct by 1900; the heath hen, an eastern relative of the western prairie chicken, was on the brink of extinction, and the great auk, a flightless bird of the northeast coast, was extinct.

Conservation Policy Framework

The policy framework that emerged by the 1930s to address such issues emphasized protection of forests from wildfire and of wildlife from overharvest. It also emphasized the management of both forests and wildlife using scientific principles. Specific actions focused on the following:

- Promoting and encouraging the protection of forests, regardless of ownership, from wildfire, insects, and disease.
- Acquiring scientific knowledge about the management of forests and wildlife, and improving the use of wood products.
- Encouraging the productive management of private forest lands through tax incentives and technical and financial assistance.
- Adopting and enforcing strong state and federal wildlife conservation laws.
- Acquiring and managing public lands for both commodity and amenity uses and values.

Public policy also focused on cooperative efforts among federal, state, and private sector interests to achieve common goals.

Rise of the Resource Professional

Despite the ambition, the lack of technically trained foresters seriously handicapped the introduction of scientific forest management early in the century. In 1900 only a handful of foresters worked in the United States; most of them had studied the European experience, including conditions that often applied poorly to American forests. Therefore one of the first steps in the scientific management of U.S. forests was to expand the number of trained forestry professionals.

In 1900 only two colleges offered forestry curricula—Cornell and Yale. By 1915 there were thirteen; and ten more were operating by the time of World War II. In 1909 only 91 foresters received bachelor's and master's degrees. That figure had risen dramatically by 1939, when 1,200 received such degrees.

U.S. forestry research and practical experience also increased, providing a sound foundation from which forestry professionals could work. By the 1950s, more and more wildlife biologists, soil scientists, hydrologists, forest engineers, and people in other natural resource disciplines were graduating from U.S. colleges.

Increased Research

Forestry research shifted as interest in the subject grew. Before 1900 forestry research focused on identification and description of trees, shrubs, and forest vegetation; timber use and consumption; and probable future timber supplies. That began to change after 1900.

In 1910 the Forest Service established the Forest Products Laboratory in Madison, Wisconsin. Its purpose was to seek ways to improve the utilization of wood products. Even before 1910, Forest Service researchers had been working with railroad companies seeking ways to extend the service life of wooden crossties through preservative treatments and other methods.

In 1915 the Forest Service created the Research Branch for scientific and technical investigations. Forestry research grew further with passage of the McSweeny-McNary Act in 1928. The act expanded forestry research and authorized regional forestry research stations and a nationwide forest inventory program. Research at forestry schools and state agricultural experiment stations also grew during the 1930s.

Following World War II, research improved and developed in the Forest Service, as well as at forestry schools and state agricultural research stations. The forest industry also stepped up its research efforts, making headway in silviculture, genetics, insect and disease control, and plantation and nursery practice.

Researchers discovered new, efficient ways to use wood and at the same time developed new products. For example, plywood soon replaced lumber for sheathing on buildings.

Fire Protection

In the first two decades of the century, wildfire ran essentially unchecked through America's forests. Natural fire has always been an important ecological factor in most North American forests, and fire was also used as a management tool by many native peoples. But the extensive logging and land clearing during the 19th century greatly increased both the extent and the destructiveness of wildfires.

Before 1930 from 20 million to 50 million acres commonly burned each year; few forest areas were effectively protected. In the 1920s there were about 80 million acres of land that were unstocked, largely because of repeated wildfires. Few if any areas were replanted after logging, at least in part because of the risk of loss to fire.

It became clear that the fire problem had to be addressed. Europe, which had a negligible fire problem, left American foresters without a model.

In 1902 a series of catastrophic fires near Yacolt, Washington, burned more than a million acres and took thirty-eight lives. These fires encouraged the forest industry to set up private fire protection associations. In 1910 devastating fires in northern Idaho and northwestern Montana caused considerable property damage and loss of life and helped galvanize federal efforts in fire control. William B. Greeley, who was in charge of the Idaho and Montana region of the Forest Service at the time of the 1910 fires and later became Forest Service chief, campaigned vigorously for stronger fire suppression programs. The fires prompted Congress in 1911 to pass the Weeks Act, which authorized federal matching funds for state fire-control agencies.

The Clarke-McNary Act in 1924 augmented cooperative federal and state fire suppression efforts as well as existing funding under the 1911 Weeks Act. This fire control system covered federal, state, and private lands in a cooperative effort.

By the end of the 1930s these programs began to show results. However, it took three decades before wildfires were reduced to present levels. Of all the efforts to educate the public about fires, the introduction of Smokey Bear as a symbol of fire prevention was perhaps the most successful and widely recognized.

By the late 1950s, as a result of increasingly sophisticated fire protection, suppression, and public education, both the area burned and the size of fires had been substantially reduced. Today only 3 million to 5 million acres burn in an average year (see figure 5).

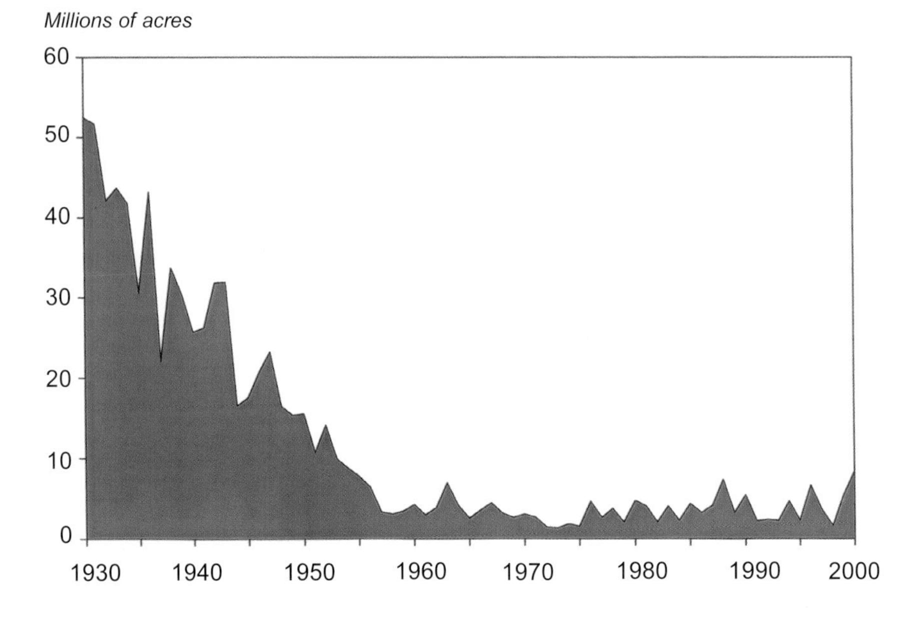

Figure 5. Wildfires scorched 40 million to 50 million acres of land each year during the 1930s, a fact that ultimately made fire control a national priority. By 1960 the area burned had been reduced by 80 to 90 percent, to 2 million to 9 million acres annually through cooperative federal, state, and local efforts in fire prevention, suppression, and public education. The buildup of forest fuels in the western United States has resulted in an increase in the area and destructiveness of wildfire in recent years.

Reducing the risk of loss to wildfire was a prerequisite to effective forest management in the United States because it reduced the risk for long-term investments in timber growing. However, the nation still loses about 6.2 billion cubic feet of timber a year to insects, diseases, and wildfire—about a third of the volume of wood the nation consumes annually.

Source: Wildland Fire Statistics, Fire and Aviation Management, USDA Forest Service.

Wildfire protection has been so effective that concerns are now being raised in many areas over the need to reintroduce fire on a controlled basis into forest ecosystems where it should play an important role in forest and ecosystem health and diversity.

State Efforts in Forest Conservation

State forestry programs preceded federal action. In 1885 California and New York established forestry commissions, and the Empire State even set aside the Adirondack Preserve to protect water supplies for the Erie Canal. However, it was not until the 1911 Weeks Act provided federal matching funds to forest fire protection agencies that state programs grew. The 1924 Clarke-McNary Act further bolstered federal support of states through a major study of forest land taxation and assistance with tree nurseries.

During the 1920s and 1930s many states reexamined their constitutions to see whether property tax changes could be made that would give special consideration to forest lands. The tax problem for forest managers was that although trees were taxed annually, they produced income only after long intervals. This situation, and the possibility that taxes might substantially increase during a managed forest's rotation, reduced incentives for reforestation following logging. Nationally, the property tax situation was modified piecemeal. Today forest land generally receives a more favorable tax treatment.

In the 1940s, with passage of various state forest practice laws, the Forest Service campaign for federal regulation was ended when states became the regulators of private forest practices. Early forest practice laws emphasized fire protection and reforestation. Recent revisions include requirements for successful reforestation and reflect broad concerns for the environment. Game also fell under state regulation, even game in national forests. In most states, fish and game agencies, funded largely by sport license fees, established bag limits and hunting seasons in an effort to enhance the wild populations. Predator controls shifted from extermination to balanced maintenance as a way to ensure long-term wildlife health.

Stabilization of Timber Consumption

One significant development in the forest conservation picture after 1900 was the stabilization of timber consumption, followed by a modest but persistent decline in the timber volume used. Per capita consumption rates for wood, which in 1905 were more than 500 board feet per year, dropped to less than 200 board feet by 1970. Even though population continued to increase, by the 1940s national wood production was about 15 percent lower than in the early 1900s.

There were various reasons for the leveling off and subsequent decline of timber consumption. One was technology. Fossil fuels replaced wood fuels, and wood substitutes, such as steel and concrete, replaced wood in structural applications. The rising real price of wood encouraged such shifts. The price of timber, adjusted for inflation, had risen steadily since 1800, increasing fivefold during the century. The real prices of competing materials were steady or declining during this period and throughout most of the 20th century as well.

After World War II, increasing real prices for wood created powerful incentives not just to use wood substitutes but also to improve the efficiency with which wood was used. Tree sizes and species formerly left behind were removed. Sawmills invested in wood-saving technologies. More efficient new products, such as plywood and various panel products, were developed.

Statistics reflect these changes in technology. In 1940 plywood accounted for less than 3 percent of U.S. production of solid wood products. By 1980 plywood's share had risen to 11 percent. Expanded use of preservative treatments also reduced the demand for wood. By 1920 virtually all crossties were being treated, and by 1960 railroad use of wood had dropped to one-fifth what it had been in 1900.

Rise in Wildlife Conservation

Beginning in the late 1800s, organized sportsmen waged a protracted and ultimately successful war against market hunting. These groups vigorously supported enforcement of game laws, self-taxation to support state game management, and acquisition of habitat reserves and management areas. Sportsmen formed the National Audubon Society out of concern over commercial plume hunting. Such organized efforts saved scores of game and nongame species from extinction (see figures 6, 7, 8, and 9).

Before 1920 the primary focus was on eliminating market hunting and establishing a strong framework for the regulation of hunting. The regulatory framework that eventually emerged included the following:

- Halt market hunting of wildlife for meat and most other products, including feathers (regulated market hunting of furbearers continued).
- Eliminate spring shooting of waterfowl and other game birds.
- Establish state regulation of resident game and nongame species.
- Prohibit hunting of songbirds, plume birds, and other migratory nongame birds; prohibit interstate commerce in wildlife products taken in violation of state law.
- Enact federal regulation of sport hunting of waterfowl and other migratory game birds.

After 1920 the emphasis on game conservation expanded from regulating to improving the art and science of wildlife management. Wildlife management became part of the curriculum at many colleges and universities, and state fish and game departments devoted to scientific wildlife management and game law enforcement were established. Before 1930 most state game departments were staffed by political appointees whose competence and tenure depended on the governor.

Increased professionalism among wildlife managers, coupled with improving habitat conditions, especially on millions of acres of abandoned farmland in the East and South, provided the foundation to reintroduce wildlife species into formerly occupied ranges.

Rise of Industrial Forestry

Until the 1920s, the forest products industry showed little interest in forest management. In fact, timber companies often sold cutover tracts for farmland or even let it revert to the counties for nonpayment of taxes. Tax codes had an effect on land use; because property taxes were based on the combined value of land and timber, landowners were implicitly encouraged to cut timber and thereby reduce their tax burden. There was little incentive for long-term investment.

By 1960 many states had changed their tax codes to base timberland taxes on bare land values, taxing the timber only upon harvest.

Modified tax codes, state laws encouraging—even regulating—fire protection and reforestation, and the rising real price of wood products prompted increased management of industrial forest lands for long-term timber growing, especially following World War II. Industrial tree planting

Trends in U.S. White-Tailed Deer Populations, 1930–1990

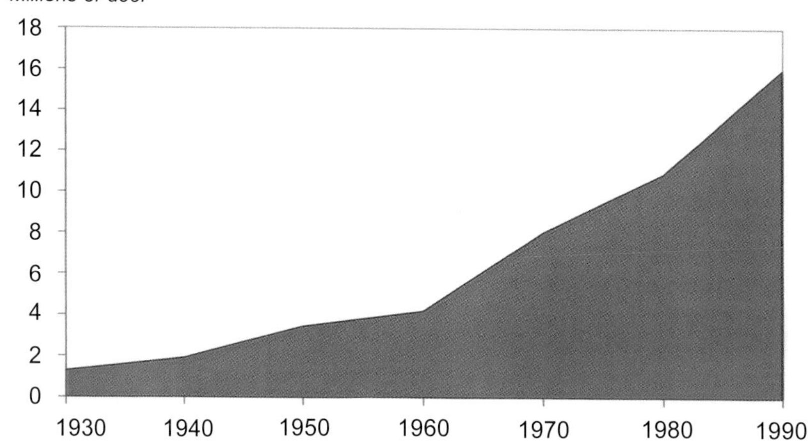

Figure 6. *Source:* Graph reflects trends, not absolute numbers; based on Chapter 8, "Wildlife," by Jack Ward Thomas, in *Natural Resources for the 21st Century*, American Forestry Association.

Trends in U.S. Wild Turkey Populations, 1900–1990

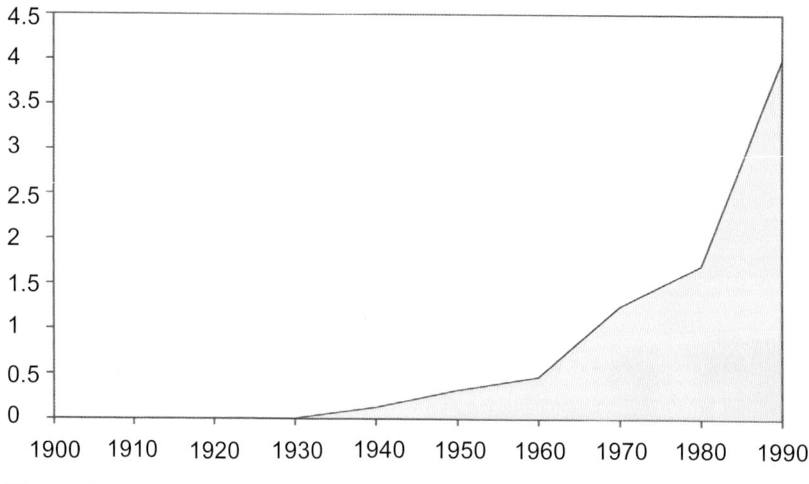

Figure 7. *Source:* National Wild Turkey Federation.

White-tailed deer, elk, pronghorn antelope, wild turkey, and many other wildlife populations, both game and nongame, have increased dramatically since 1930. These increases are the result of effective hunting laws, increases in habitat acreage of managed forests, the adaptation of species to a variety of forest conditions, and the dedicated work of federal and state wildlife agencies and private groups,

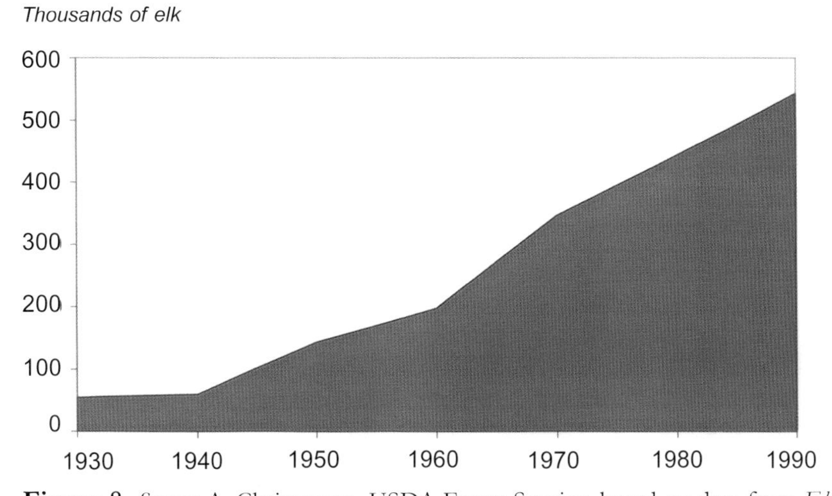

Figure 8. *Source:* A. Christensen, USDA Forest Service, based on data from *Elk of North America*.

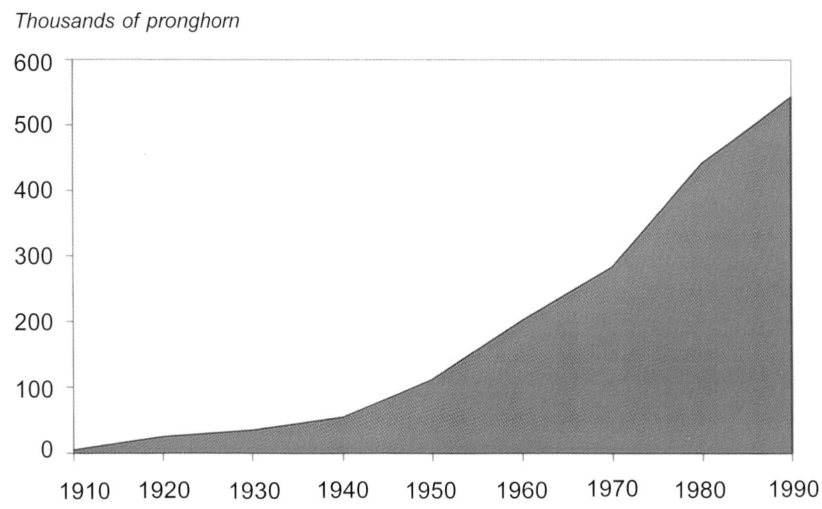

Figure 9. *Source:* Graph reflects trends, not absolute numbers; based on Chapter 8, "Wildlife," by Jack Ward Thomas, in *Natural Resources for the 21st Century*, American Forestry Association.

such as the National Wild Turkey Federation. During the past three decades there has been growing concern over some wildlife species that need specialized habitats, such as the northern spotted owl on the Pacific Coast, the red-cockaded woodpecker in the South, and some neotropical birds.

rates increased dramatically after 1950, rising from an average of about 7,000 acres a year just prior to 1945 to 1.2 million acres during the 1980s (see figure 10). Much of this tree planting was in the South.

The forest industry also began to increase forest land holdings after World War II. Between 1952 and 1987, industrial land increased by 11.6 million acres, half of which was in the South. Today about 71 percent of U.S. productive forest land is privately held. These lands provide 89 percent of the nation's timber harvest volume. Forest industry lands constitute 13 percent of the U.S. productive forest land base, yet they provide 30 percent of the nation's timber harvest volume and 37 percent of the nation's softwood timber harvest volume (see figure 11).

Stabilization of Forest Area

By the 1920s, the 300-year loss of forest land in the United States had nearly halted. Today the country actually has about the same area of forest as it did in 1920. The primary reason for forest land stabilization was the stabilization of the nation's cropland area.

Around 1920, for the first time in American history, increases in the area of cleared farmland abruptly stopped, rather than rising at the rate of population growth. Farm clearing of forests continued after 1920 in some areas, but it was offset by farmland abandonment and reversion to forest in other areas (see figure 2, p. 15).

Cropland stabilized primarily for two reasons. First, rapidly increasing numbers of motor vehicles and farm tractors made it unnecessary to continue raising large numbers of draft animals. In 1910, 27 percent of all cropland was devoted to growing food for draft animals. By 1950, the number of draft animals had dropped so dramatically that the equivalent of 70 million acres had been released to grow crops for human consumption. The second reason for the stabilization of cropland was that after 1935, spurred by the development of genetically improved hybrid crops and expanded use of chemical fertilizers and liming, agricultural productivity improved. Today American farmers commonly produce five or more times the crop yield per acre that they did in 1920.

The Eastern Forest Comes Back

Although the United States has about the same aggregate area of forest as it did in 1920, some areas have considerably more and some have less.

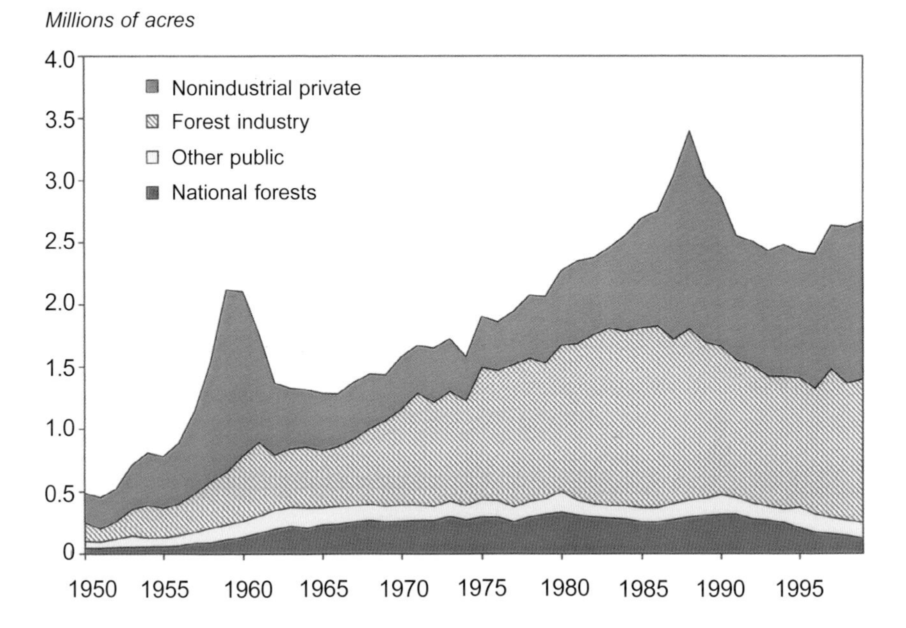

Figure 10. Tree planting increased steadily following World War II, peaking in 1988. In the 1990s, more than 25 million acres were planted. In recent years, tree planting has averaged between 2.4 million and 2.6 million acres annually. In the 1990s, more than 400 trees were planted for every child born in the United States. The peaks in planting in the late 1950s and 1980s correspond to the efforts of the Soil Bank Program and the Conservation Reserve Program, respectively, and point to the effects that government subsidies can have on acres planted.

Source: R. J. Moulton. Tree Planters Notes, USDA Forest Service, 1999.

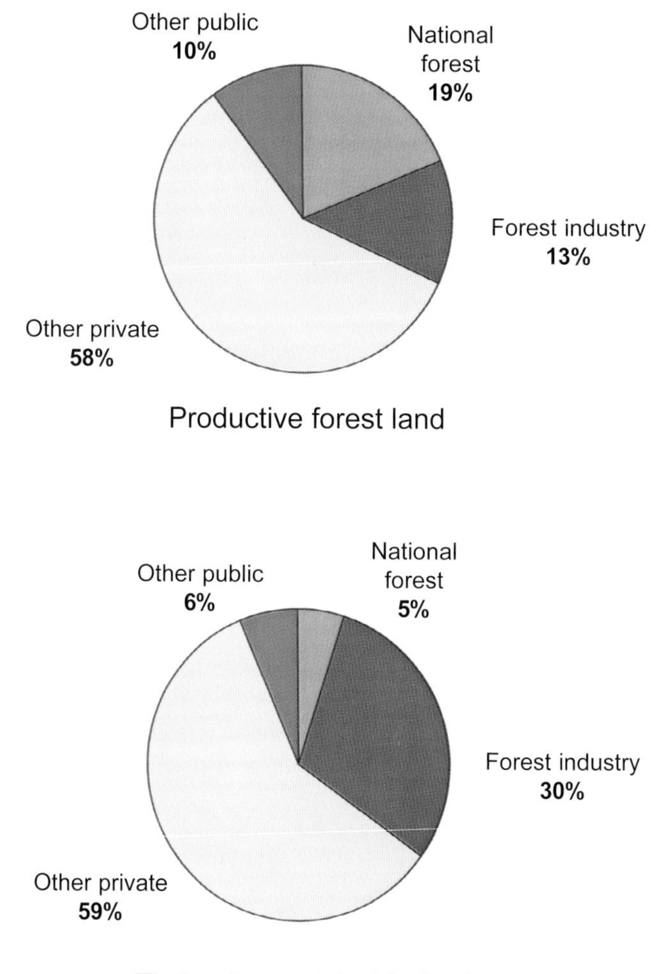

Figure 11. Just as they had encouraged improved wood utilization, increasing timber prices spurred private sector investments in timber growing, although investments other than for fire protection were not significant until after World War II, when industrial forest lands began to be managed in earnest for tree growing. Today private forests ("Forest industry" and "Other private") constitute 71 percent of U.S. productive forest land yet supply 89 percent of the wood volume harvested. The forest industry holds about 13 percent of the nation's productive forest land yet provides 30 percent of the timber harvested.

Sources: Forest Resources of the U.S., 1997. GTR-NC-219, USDA Forest Service, 2001.

Beginning in the mid-1800s, marginal agricultural land in the East and South was gradually abandoned as more productive farmlands in the Midwest were developed; the abandoned farmland often grew back as forest (see figure 12). This reversion to forest has not been generally recognized by the public.

The reasons for reversion to forest include two related factors working in concert. The first was the growth of cities, which accelerated the transition of U.S. agriculture from subsistence to commercial. The second was the nation's progressively improving transportation system, which opened up more productive western lands that could supply the growing cities. The steep slopes, small fields, and less productive lands of the East and southern Appalachians could not compete with lands of the Ohio Valley and the rest of the Midwest. The opening of the Erie Canal in 1825 was the first major step in this reversion to forest that occurred in the Northeast. Vermont is typical of the abandonment: in the 1850s, only about 35 percent of Vermont was forest, with the remainder primarily crops and pasture. By 1980, however, 75 percent of the state had become forest.

As surely as the Erie Canal and the railroads created a prosperous Midwest, they signaled the demise of agriculture in New England. The agricultural land abandonment that started in the Northeast in the 1850s gradually spread during the next century to other parts of the East, to the South, and eventually even to less productive farmlands of the Midwest. It culminated in massive farm abandonments during the Great Depression of the 1930s.

In many ways, the forest and farm landscape of the Appalachians, as well as many other parts of the East and South, has come full circle. By the 1960s and 1970s, the pattern of forest, fields, and pastures was similar to that prior to 1800, its appearance much as it must have been before the American Revolution.

The Eastern National Forests

By 1915 national forests of the West had been established in the form they retain today. These national forests, which included 162 million acres in 1915, were essentially carved out of the public domain. At that time there were no federal forests in the East because the public domain had been transferred to private ownership before the conservation movement began.

The impetus for eastern national forests had two sources: some groups advocated federal acquisition to provide general protection for cutover lands, and other groups focused on the need for flood prevention. These parallel interests converged to influence passage of the 1911 Weeks Act, authorizing

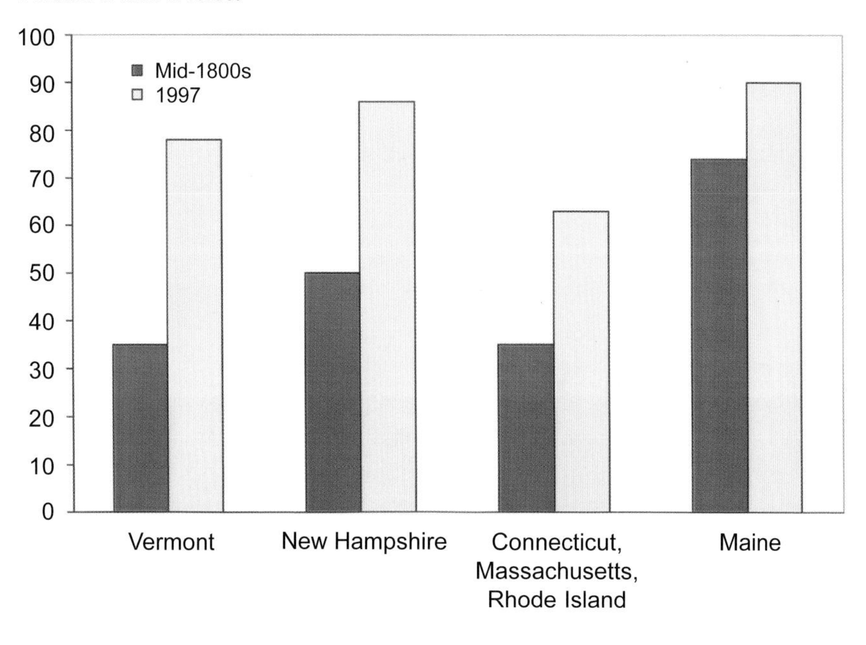

Figure 12. The amount of forest land in many parts of the East and South has actually increased by tens of millions of acres since 1900. Virtually every state east of the Mississippi has seen an increase in forest land since 1900, although the specific amounts and timing vary by state. The reasons for reversion to forest are complex. The growth of the cities accelerated the transition of U.S. agriculture from subsistence to commercial. At the same time, the nation's progressively improving transportation system opened up more productive western lands to provide for the growing cities. With their steep slopes, small fields, and less productive soils, farmers of the East and southern Appalachians were unable to compete commercially with farmers in the Ohio Valley and much of the rest of the Midwest. The process of farmland reversion to forest was greatly accelerated by the Great Depression.

Sources: Barrett, Regional Silviculture of the U.S., 1980; and Forest Resources of the U.S., 1997. GTR-NC-219, USDA Forest Service, 2001.

Top: A Harvard Forest diorama shows a hardwood-conifer forest in 18th-century New England. Middle: A century later, the forest had been largely cleared for farms. Within a few decades competition from agriculture farther west would cause this farm to be abandoned. Above: Yet another century later the farm had reverted to forest, which is now managed for timber production. *Source:* Harvard Forest Diorama, Fisher Museum, Petersham, Massachusetts. M.H. Zimmerman photo.

the acquisition of federal lands to protect the watersheds of navigable streams.

The first acquisitions under the Weeks Act were in the southern Appalachians and in the White Mountains of New Hampshire. By 1925 land had been purchased to establish the national forests today known as the Monongahela in West Virginia; the Pisgah in North Carolina; the George Washington and Jefferson in Virginia; the White Mountain in New Hampshire; the Nantahala in North Carolina, South Carolina, and Georgia; the William B. Bankhead in Alabama; the Cherokee in Tennessee; and the Allegheny in Pennsylvania.

The major acquisition of eastern national forests occurred during the Great Depression. At that time twenty-six national forests were established, ranging from the Ocala in Florida to the Nicolet in Wisconsin; from the Green Mountain in Vermont to the Mark Twain in Missouri.

By 1945, when acquisition of national forest land in the East substantially slowed, about 24 million acres of depleted farmland and cutover and burned woodlands had been incorporated into the eastern national forest system and placed under long-term management.

Postwar Demands on the National Forests

The period after World War II ushered in a substantial increase in demand for a variety of forest products as well as nontimber uses and values. Prior to the late 1940s, management of national forests was generally custodial or focused on meeting demands for resources in the surrounding area. After the war, as millions of GIs returned home and started families, demand for timber to use in housing increased dramatically, and the nation looked to the national forests to meet that demand. The roads into national forests had improved by the late 1940s and many of the more accessible private lands had been logged to provide timber for the war effort.

National forest timber sales increased from about 3 billion board feet in the late 1940s to about 11 billion board feet by the early 1960s. By the 1960s, national forests met about one-sixth of the nation's total consumption of wood volume, and a quarter of its softwood sawtimber needs, a primary source of lumber and plywood for housing.

This increase not only met the critical need for timber, it also took pressure off private forest lands, many of which had been heavily used to meet the needs of the war effort. Standing inventory was affected by this demand, as was regrowth (see figures 13 and 14). Balancing harvest with growth in

Trends in Average Net Growth per Acre by Ownership, 1952–1996

Cubic feet per acre per year

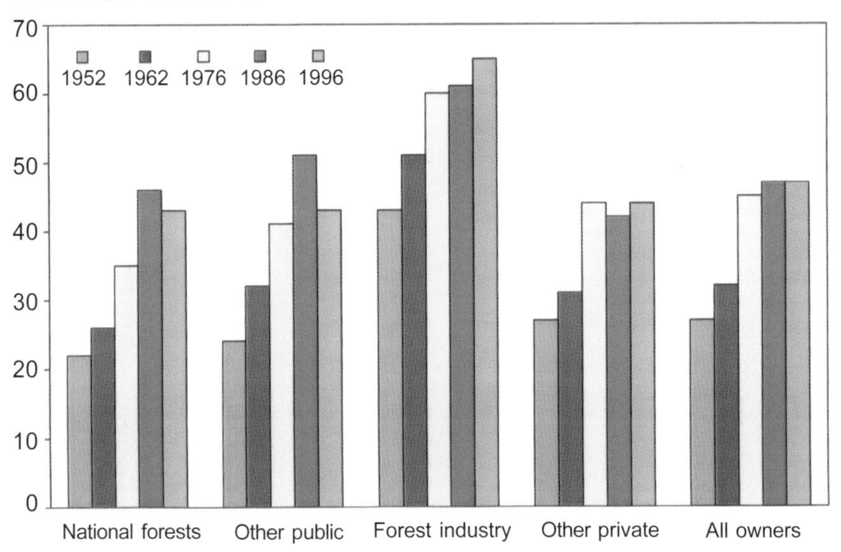

Figure 13. Forest growth has increased dramatically on all ownerships over the past several decades. Since 1952, net annual growth for all U.S. forests has increased 69 percent.

Sources: Forest Resources of the U.S., 1997. GTR-NC-219, USDA Forest Service, 2001.

Trends in U.S. Standing Timber Volume per Acre by Region, 1953–1997

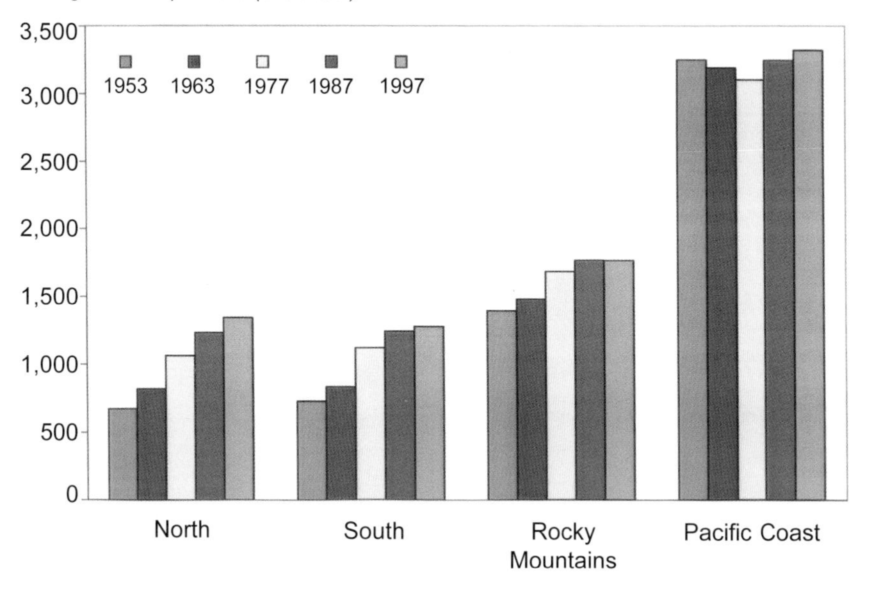

Figure 14. Since 1953, the amount of timber standing in all U.S. forests has increased dramatically in all regions except the Pacific Coast, where per-acre volume declined slightly until recently because of the harvest of old-growth timber, but is now greater than 1953 levels. Timber volume per acre has increased by 37 percent nationally since 1953. In the eastern United States, it has almost doubled.

Sources: Forest Resources of the U.S., 1997. GTR-NC-219, USDA Forest Service, 2001.

a system of multiple owners, and the transition from old-growth to second growth, proved challenging (see figure 15).

The 1950s also witnessed a substantial increase in demand for other national forest uses and values. An increasingly mobile and affluent population began to look to these lands for outdoor recreation. National forest recreational visits increased from 18 million in 1946 to 93 million in 1960 (see figure 16).

The increased demands on national forests led to new laws in the 1960s. The Multiple Use–Sustained Yield Act of 1960 required that national forests be managed for a variety of uses and values, including outdoor recreation, wildlife, timber, grazing, and watershed protection. In effect, this law reflected the uses and management already occurring on these lands.

The Wilderness Act, passed in 1964 after much debate, provided for the preservation of significant areas of national forest land in their natural and untrammeled condition. Timber sales and most other commodity uses were prohibited in these areas. By 1990 more than 33 million national forest acres had been designated as wilderness. Approximately half of this land is forested.

In 1974 the Renewable Natural Resources Planning Act (RPA) required the Forest Service to carry out periodic assessments of the national long-term demand and supply situation for all renewable resources and to lay out a policy and programmatic framework for addressing projected resource demands and needs. In 1976 the National Forest Management Act provided detailed guidelines for national forest land management and for public participation in national forest decision-making. This last statute clearly reflected a change in congressional thought; instead of having broad mandates, the agency would operate under more detailed guidelines.

U.S. Forests Today

It is a measure both of the inherent resilience of our forests, and of the success of the policies put in place in response to public concerns in the early decades of the 20th century, that forest conditions over much of the United States have improved dramatically since 1900:

- Forest land area has stabilized.
- Acreage of uncontrolled forest fires is down 90 percent.
- Forest growth now exceeds harvest.
- Average timber volume per acre is up one-third since 1952; in the East, it has almost doubled.

U.S. Timber Growth and Removals, 1920–1997

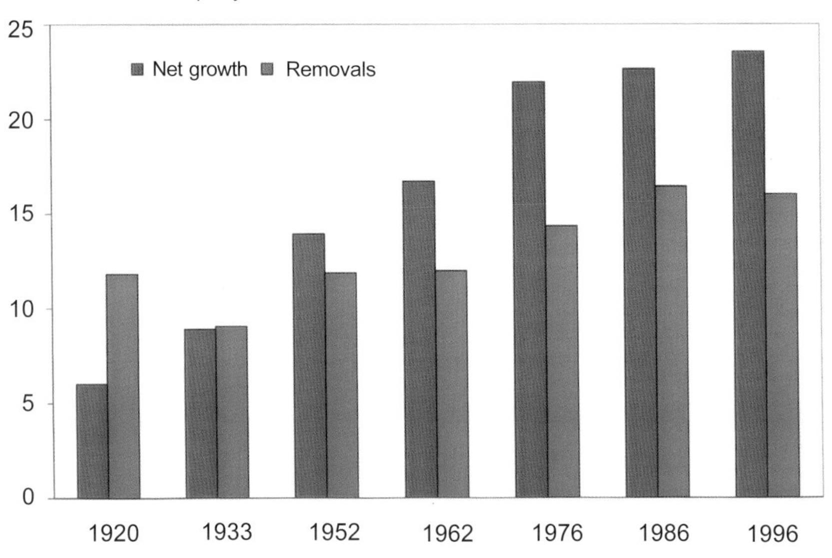

Figure 15. In 1920, timber harvest rates nationally were double the rate of forest growth, but by 1952, net annual growth had exceeded annual harvest from all U.S. forests. By 1996, net annual growth was 3.9 times what it had been in 1920. In 1996, net growth exceeded harvest by 47 percent, or 7.5 billion cubic feet (about 24 billion board feet).

Sources: Forest Resources of the U.S., 1997. GTR-NC-219, USDA Forest Service, 2001; A National Plan for American Forestry (Copeland Report), Senate, 73rd Cong., 1st sess., Report on S. Res. 175, S. Doc. 12. Washington, D.C.: GPO, 1933; and Timber Depletion, Lumber Prices, and Concentration of Timber Ownership (Capper Report), Senate, 66th Cong., 2nd sess., report on S. Res. 311, Washington, D.C.: GPO, 1920.

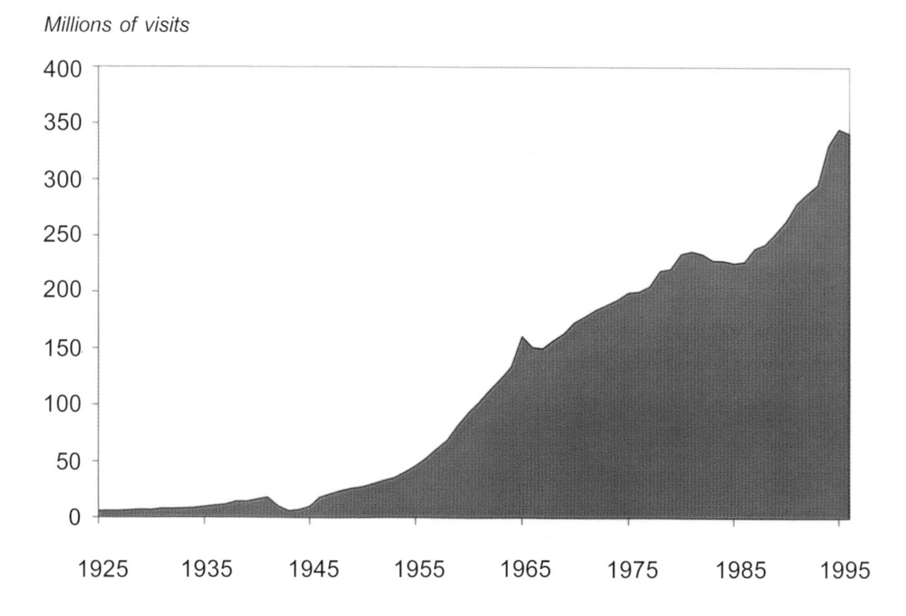

Figure 16. After World War II, steadily increasing national economic growth provided the basis for increasing personal income and leisure time. That, along with the proliferation of automobiles, revolutionized the recreational habits of the American people. Growth in recreational visits to the national forests reflects a pattern common to other public lands as well.

Increased recreational demands came at the same time that the nation's public forests were also experiencing increased demands for other uses. Such pressures have been felt especially in the past three decades as conflict over the use and management of public lands has intensified.

Source: USDA Forest Service Recreation Division, Washington, D.C.

- Reforestation is at record levels.
- Eastern forests have staged a comeback.
- Recreational use is at an all-time high.
- Wood is used with greater efficiency.

Forest Wildlife Today

Several species of American wildlife became extinct as a result of forest changes and human uses during the 20th century, including the passenger pigeon, heath hen, and Carolina parakeet. An even larger number of subspecies and wildlife populations were substantially diminished.

Many of those species that were threatened with extinction in 1900, however, have come back in abundance. Thanks to actions begun in the early decades of the 20th century, today most forest wildlife species are both more abundant and more widespread than they were in 1900. Many species that would have been on an endangered species list, had one existed a century ago, are now abundant. Examples include wild turkey; beaver; egrets, herons, and many other wading birds; many species of shorebirds; wood ducks and several other species of ducks; whistling swans; Rocky Mountain elk, pronghorn, bighorn sheep, black bear; even white-tailed deer throughout most of its range. Many other species, although not actually on the brink of extinction in 1900, are today both more abundant and more widespread than they were in 1900. Since the 1930s, forest wildlife that can tolerate a relatively broad range of conditions (so-called habitat generalists) has increased. Most American forest wildlife species are habitat generalists, perhaps because the natural dynamics of North American forests cause frequent disturbances in the natural regime (see figures 6, 7, 8, and 9).

Some species abundant in forests prior to European settlement—particularly large predators and herbivores, such as wolves, elk, and bison, which need large home ranges—have not returned to large areas where they formerly were common. Yet even many of these species have come back in areas large enough to accommodate their needs for a home range. But although there have been many successes, problems remain. Some species with specialized habitat requirements are of concern today. Examples include the following:

- The red-cockaded woodpecker and gopher tortoise, both natives of fire-created southern pine savannas and woodlands.

- The Kirtland's warbler, which lives in young jack pine forests of Michigan.
- The northern spotted owl, which occupies mature and old-growth forests in the West.

Some forest wildlife species—for example, Kirtland's warbler—require active management of young forests for their survival. Many other species, including a wide variety of game and of nongame species, need a mixture of forest and forest edge environments. Some, like grizzly bears, wolves, elk, and forest-interior birds, need large, contiguous areas of habitat. Some require old and ecologically diverse forests. Others, like the red-cockaded woodpecker, need both mature forest and other specific habitat conditions, such as open savannas and woodlands created by frequent ground fires. Even the old-growth Douglas-fir forests in which the northern spotted owl lives are subclimax forest types that will eventually move toward different forest conditions unless there are occasional, stand-replacing wildfires.

Threats to Forests Today

An objective evaluation of the performance of conservation policies and practices since 1900 suggests some impressive gains. Nevertheless, some environmental trends are not positive, and much work remains to be done. Problems include habitat fragmentation due to residential subdivision and urban development; loss and deterioration of the forest and grassland habitats that had been created and maintained by frequent, low-intensity fire; reduction and fragmentation of late successional and old-growth forest habitats due to timber harvesting; loss and degradation of riparian and wetland habitats; and effects of air pollution on forests in some areas, to name a few. Of particular concern are rare and unique ecosystem types and the species with specialized habitat requirements that are associated with them.

One significant general threat is from introduced exotic plants, animals, and diseases. There is a long history of damage to forests from introduced biological agents, including white pine blister rust, chestnut blight, Dutch elm disease, gypsy moth, and more recently, hemlock woolly adelgid and the Asian long-horned beetle. Increasing world trade in forest products, and international trade generally, offers greater opportunity for such introductions. Introduced exotic animals also pose a significant threat to native wildlife species, which they may displace and outcompete.

The Forest in a Broader Context

The American Forestry Association, formed in 1875, and the Sierra Club, formed in 1892, are both tangible examples of public concern for the forested environment. Other concerned groups included the Boone and Crockett Club (1888), National Audubon Society (1905), and Izaak Walton League (1922). The creation of the Wilderness Society in 1935 reflected a deepening sense that some land should remain relatively undisturbed, but not until the 1960s did this segment of the public begin to exercise fully its clout in setting public land priorities.

At that time of turmoil, when many of society's institutions were severely challenged, the modern environmental movement took form, moving beyond merely advocating wilderness preserves. Environmental quality became a high priority; Earth Day, a public celebration, followed closely on the heels of the National Environmental Policy Act, a federal watershed in managing lands and resources. Litigation became a weapon as public organizations made full use of new statutes. The National Environmental Policy Act and other statutes mandated public involvement in land management, which included forest land.

As forest conservation practices set in place decades before began to work, and the nation demonstrated its ability to meet wood product needs from private and public lands, more forest lands have been set aside for parks, wilderness areas, and similar designations under which timber removal is prohibited. The area of such setasides has increased significantly in recent years. Today about 52 million acres of forest has been reserved. The area, the size of Kansas, is about double the setaside acreage of the 1950s (see figure 17).

American Forests: A Transformed Heritage

Today our forests represent a substantially transformed legacy—certainly in comparison with 1600. But our forests have also been substantially transformed since 1900, a dimension not commonly understood.

Attitudes about the nation's forests have changed profoundly over the years. Native peoples viewed the forest in a spiritual context, but they also took a utilitarian approach and managed forests to serve their own ends. European Americans initially viewed forests both as an encumbrance to agriculture and as an inexhaustible resource. At first they used the forest—its wildlife, wood products, and land—to meet their subsistence needs for food and energy, much as the native population had done.

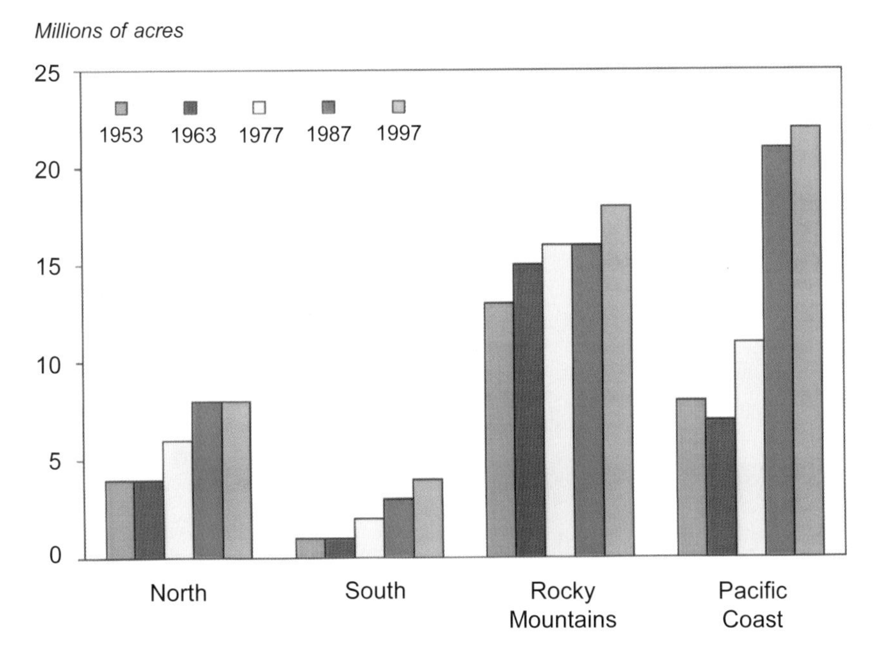

Figure 17. As personal incomes increased and the nation's population became more urbanized and mobile, interest grew in setting aside land in parks, recreation areas, and reserves. In addition, the success of forest conservation practices began to demonstrate the nation's ability to meet increasing wood product needs from both private and public lands.

Consequently, there has been a significant increase in the area of productive forest land set aside for amenity values in parks, wilderness areas, and similar designations under which timber harvest is prohibited.

Currently about 52 million acres of forest land has been so designated—about double what was set aside in 1953. This is an area the size of Kansas. Withdrawals have centered in the western United States, where the federal government is the largest forest landowner.

Sources: Forest Resources of the United States, 1997; GTR-NC-219. USDA Forest Service, 2001.

Later, the abundant wealth of the forests built the homes, cities, and transportation infrastructure of a growing nation. Lands previously occupied by forests were used to feed a rapidly increasing population.

Scarcely more than a century ago, it became increasingly clear that old approaches were not sustainable. Americans began to view forests and wildlife not as products to be mined or foraged, but as resources that could be managed scientifically over the long term, yielding products and services without unduly disrupting the basic resource. We called this idea conservation.

As the nation's population has continued to urbanize, the principle of forest conservation for products and services has remained, but its role and scope have expanded. A few decades ago Americans started to view forests as attractive settings for outdoor recreation and as places for human spiritual renewal. Recently this view has evolved to a view of forests as ecosystems that support a complex web of life, of which humans are a part (see figure 18).

Although it is impossible to predict how the American view of forests may change in the future, the past provides information about how these forests came to be what they are today.

Lessons of the Past and Challenges for the Future

The U.S. population has more than tripled since 1900, and the standard of living is substantially higher. At the same time our forests and wildlife are, in most of their major dimensions, in significantly better condition today than they were a century ago.

American forests and wildlife have demonstrated a resilience and responsiveness to management undreamed of by conservationists at the turn of the last century. These leaders were almost universally pessimistic about the future. Forest Service Chief Gifford Pinchot and others predicted a timber famine coupled with significantly increased wood product prices and consequent economic hardship. Wildlife leaders like William T. Hornaday foresaw the imminent extinction of scores of species.

The timber famine never came; most species whose extinction was prophesied have since recovered and many are abundant today. Predictions by these early conservationists reflected what they felt was likely to occur if trends continued. Their words were a call-to-arms. Action was taken: new policies were debated and implemented. History has demonstrated that these policies, coupled with the natural resilience of the resource, have caused forests and their wildlife to come back.

Trends in U.S. Population Growth by Rural and Urban, 1790–1990

Millions of people

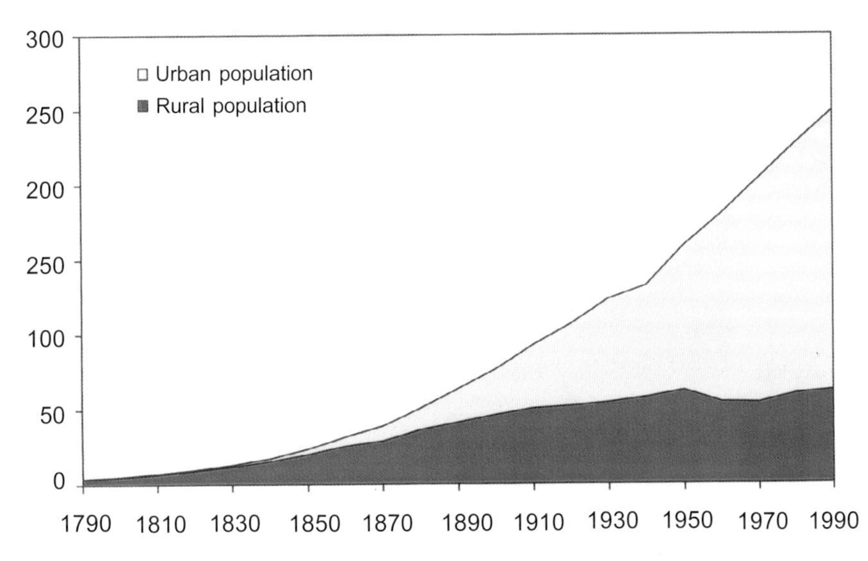

Figure 18. One of the most profound changes in American society in the 20th century was its transition from a rural, agrarian society to an urban, industrialized nation. This change was accompanied by a corresponding physical and psychological separation of its people from the land and resources that sustain them.

In a world of farms, forests, and small towns, the linkages between food and fields and between forests and home and hearth were clear and sustained by personal experience. In a world of cities and suburbs, of offices and air-conditioning, these linkages have become more obscure, and for many people, virtually nonexistent. Yet today's urbanized society is no less dependent upon the products of its forests and fields than were the subsistence farmers of America's past.

Sources: Statistical Abstract of the U.S., 2000; and Historical Statistics of the United States from Colonial Times to 1970; both from Bureau of Census, U.S. Department of Commerce, Washington, D.C.

In addition to policy, certain actions and conditions unforeseen by early conservation leaders have also been important to the improved condition of forest and wildlife resources. One, of course, was resiliency; even Forest Service projections have consistently underestimated forest growth. Wildlife specialists have also been surprised at the recovery rates of many species, once placed under protection and management.

Another action was consumers' conversion from wood energy to fossil fuels, relieving American forests of that burden as population grew. Indirectly, use of fossil fuels in internal combustion engines substantially reduced pressure to clear forest land for agriculture because it released millions of acres of cropland to grow food for humans rather than for draft animals. Petroleum was also the base for fertilizers and pesticides that substantially increased agricultural productivity after 1930.

American forests have been principal beneficiaries of the remarkable improvement in agricultural productivity over the second part of the 20th century. The seemingly inexorable, three-century-long conversion of U.S. forests to farmland largely halted in the 1920s. Today about the same area is devoted to cropland as in 1920, despite a doubling of the nation's population. On top of this, U.S. farmers feed, through exports, the equivalent of more than 100 million people throughout the world.

Finally, a factor unrecognized by early conservation leaders was the effect that increasing prices for wood products would have in encouraging reduced consumption and increased supply. Real price increases for wood created incentives for efficient use, including less left behind after logging, better utilization by sawmills, and more efficient use in end-product applications through improved engineering, protection from rot through preservative treatment, and similar measures. Price increases also encouraged use of substitutes for wood, such as steel and concrete. These market responses were the primary reason that wood consumption did not rise after 1910 as it had in previous decades. Projections of impending shortages were based on assumptions that such past trends would continue.

But there are still significant issues and controversies surrounding management of U.S. forests. In recent years the growing urbanization, affluence, and mobility of Americans have caused a virtual revolution in the expectations and demands that the public places on forests. Some of these demands are in direct conflict with traditional forest values and uses.

In the last decade, the debate between people advocating the use and management of forests for commodity products and people wanting to minimize human influences and emphasize amenity values (particularly on public forests) has become increasingly shrill and divisive. Utilitarian use of the forest for commodities is often viewed as irreconcilably in conflict with its protection for amenity and natural values—and on a personal level, or when the focus is on an individual parcel of land, these goals frequently are. Yet in a larger sense and scale, they are not only compatible; indeed, they are inextricably linked. It is a measure of the success of its past conservation policies that the United States has the forest abundance that allows it the capacity—unique to only a handful of other nations—to consider such choices.

But as always, there are limits to such choices. Society remains dependent on forests for a wide variety of economic products. Indeed, the United States consumes more wood products today than at any time in its history. Today Americans use about as much wood on a tonnage basis as the combined total for most other raw materials, such as steel, plastics, aluminum, other metals, and cements.

As human population increases and demands on natural resources grow, the challenge for society and its land managers is to find ways to realize both commodity products and amenity values from the same area of forest. This increasingly must become the dual focus for the concept of land stewardship and forest sustainability.

Suggested Reading

American Forestry Association. *Natural Resources for the 21st Century*. Washington, D.C.: Island Press, 1990.

Botkin, D. *Discordant Harmonies: A New Ecology for the Twenty-First Century*. New York: Oxford University Press, 1990.

Clawson, M. Forests in the Long Sweep of American History. *Science* 204 (1979): 1168–74.

Clepper, H., and Meyer, A.B. *American Forestry: Six Decades of Growth*. Washington, D.C.: Society of American Foresters, 1960.

Cronon, W. *Changes in the Land: Indians, Colonists, and the Ecology of New England*. New York: Hill and Wang, 1985.

________. *Nature's Metropolis: Chicago and the Great West*. New York: W. W. Norton and Company, 1991.

Crosson, P. Cropland and Soils. Discussion Paper ENR90-03. Washington, D.C.: Resources for the Future, 1989.

Denevan, W.M. The Pristine Myth: The Landscape of the Americas in 1492. *Annals of the Association of American Geographers* 82 (no. 3, 1992).

Doolittle, W.E. Agriculture in North America on the Eve of Contact: A Reassessment. *Annals of the Association of American Geographers* 82 (no. 3, 1992).

Fedkiw, J. The Evolving Use and Management of the Nation's Forests, Grasslands, Croplands, and Related Resources. USDA-Forest Service, General Technical Report RM-175. September 1989.

Frederick, K.D., and Sedjo, R.A. eds. *America's Renewable Resources: Historical Trends and Current Challenges*. Washington, D.C.: Resources for the Future, 1991.

Gruell, G.E. Fire and Vegetative Trends in the Northern Rockies: Interpretations from 1871–1982 Photographs. Intermountain Forest and Range Experiment Station, General Technical Report INT-158, USDA-Forest Service. Ogden, Utah, December 1983.

Harper, R.M. Changes in the Forest Area of New England in Three Centuries. *Journal of Forestry* 16 (January 1918): 442–52.

Olson, S.H. *The Depletion Myth: A History of Railroad Use of Timber*. Cambridge, Massachusetts: Harvard University Press, 1971.

Perlin, J. *A Forest Journey: The Role of Wood in the Development of Civilization*. New York: W.W. Norton and Company, 1989.

Raup, H.M. The View from John Sanderson's Farm: A Perspective for the Use of Land. *Forest History* 10 (April 1966): 2–11.

Reiger, J.F. *American Sportsmen and the Origins of Conservation.* Norman: Oklahoma University Press, 1986.

Sedjo, R.A. The Nation's Forest Resources. Discussion Paper ENR 90-07, Washington, D.C.: Resources for the Future, January 1990.

Shands, W.E. The Lands Nobody Wanted: The Legacy of the Eastern National Forests, in *The Origins of the National Forests*, H.K. Steen, ed. Durham, NC: Forest History Society, 1992.

Smith, B.D. Origins of Agriculture in Eastern North America. *Science* 246 (1989): 1566–71.

Steen, H.K. *The U.S. Forest Service: A History*. Seattle: University of Washington Press, 1976.

Thompson, D.Q., and Smith, R.H. The Forest Primeval in the Northeast—A Great Myth? *Proceedings of the Annual Tall Timbers Fire Ecology Conference* 10 (1970): 255–65.

Trefethen, J.B. *An American Crusade for Wildlife*. New York: Winchester Press and the Boone and Crockett Club, 1975.

U.S. Bureau of the Census. Historical Statistics of the United States from Colonial Times to 1970, Bicentennial Edition, Part 1. Washington, D.C.: U.S. Department of Commerce, 1975.

U.S. Department of Agriculture. Analysis of the Timber Situation in the United States, 1952–2030. Forest Resources Report No. 23. USDA-Forest Service, 1982.

________. Forest Resources of the U.S., 1992. General Technical Report RM-234. USDA-Forest Service, September 1993.

________. Forest Statistics of the United States, 1987. Resource Bulletin PNW-RB-168. USDA-Forest Service, 1989.

Van Lear, D.H., and Waldrop, T.A. History, Uses, and Effects of Fire in the Appalachians. Southeastern Forest Experiment Station, USDA-Forest Service, General Technical Report SE-54, 1989.

Whitney, G.G. and Davis, W.C. From Primitive Woods to Cultivated Woodlots: Thoreau and the Forest History of Concord, Massachusetts. *Journal of Forest History* 30 (October 1986): 70–81.

Williams, Michael. *Americans and Their Forests: A Historical Geography*. New York: Cambridge University Press, 1989.

About the Author

Douglas W. MacCleery is a professional forester who has worked in natural resource management and policy for his entire career. He has bachelor's and master's degrees in forestry from Michigan State University.

MacCleery worked in northern California for seven years as a field forester for the USDA-Forest Service on the Tahoe and Shasta-Trinity National Forests. In this job he assisted in the multiple purpose management of the national forests for timber production, recreation, wildlife management, livestock grazing, and watershed production.

He then left the Forest Service for experience in the private sector, taking a position as a forest policy analyst for the National Forest Products Association in Washington, D.C.

Between 1981 and 1987, MacCleery was deputy assistant secretary for natural resources and the environment in the U.S. Department of Agriculture, a position in which he had oversight of the Forest Service and Soil Conservation Service.

In 1987 MacCleery returned to the Forest Service in Washington, D.C. His current position is Senior Policy Analyst in the Forest and Rangelands Management Staff.

For the past several years, MacCleery has been compiling a history of how U.S. forests have changed from precolonial times to the present. The focus is on how the relationship between humans and their forests has evolved over the years, on how the deteriorating forest and wildlife situation at the end of the 19th century led to the first national environmental movement, on the policies that emerged from that movement to address forest and wildlife depletion, and on how the forest and wildlife situation has changed since 1900 in response to those policies and to other factors.

The idea behind the history, which focuses on forests, agriculture, and wildlife, is that informed choices about the future of our forests and wildlife should be based in part on knowledge of how they came to be what they are today.